A RISK REDUCTION STRATEGY FOR HUMAN EXPLORATION OF SPACE

A Review of NASA's Bioastronautics Roadmap

Committee on Review of NASA's Bioastronautics Roadmap

Board on Health Sciences Policy

Space Studies Board

Aeronautics and Space Engineering Board

David E. Longnecker and Ricardo A. Molins, *Editors*

INSTITUTE OF MEDICINE *AND*
NATIONAL RESEARCH COUNCIL
OF THE NATIONAL ACADEMIES

THE NATIONAL ACADEMIES PRESS
Washington, D.C.
www.nap.edu

THE NATIONAL ACADEMIES PRESS 500 Fifth Street, N.W. Washington, DC 20001

NOTICE: The project that is the subject of this report was approved by the Governing Board of the National Research Council, whose members are drawn from the councils of the National Academy of Sciences, the National Academy of Engineering, and the Institute of Medicine. The members of the committee responsible for the report were chosen for their special competences and with regard for appropriate balance.

This study was supported by Award No. NASW-02031 between the National Academy of Sciences and the National Aeronautics and Space Administration. Any opinions, findings, or conclusions expressed in this publication are those of the author(s) and do not necessarily reflect the view of the organizations or agencies that provided support for this project.

International Standard Book Number 0-309-09948-X

Additional copies of this report are available from the National Academies Press, 500 Fifth Street, N.W., Lockbox 285, Washington, DC 20055; (800) 624-6242 or (202) 334-3313 (in the Washington metropolitan area); Internet, http://www.nap.edu.

For more information about the Institute of Medicine, visit the IOM home page at **www.iom.edu.**

Cover credit: NASA/John Frassanito and Associates

Printed in the United States of America.

THE NATIONAL ACADEMIES

Advisers to the Nation on Science, Engineering, and Medicine

The **National Academy of Sciences** is a private, nonprofit, self-perpetuating society of distinguished scholars engaged in scientific and engineering research, dedicated to the furtherance of science and technology and to their use for the general welfare. Upon the authority of the charter granted to it by the Congress in 1863, the Academy has a mandate that requires it to advise the federal government on scientific and technical matters. Dr. Ralph J. Cicerone is president of the National Academy of Sciences.

The **National Academy of Engineering** was established in 1964, under the charter of the National Academy of Sciences, as a parallel organization of outstanding engineers. It is autonomous in its administration and in the selection of its members, sharing with the National Academy of Sciences the responsibility for advising the federal government. The National Academy of Engineering also sponsors engineering programs aimed at meeting national needs, encourages education and research, and recognizes the superior achievements of engineers. Dr. Wm. A. Wulf is president of the National Academy of Engineering.

The **Institute of Medicine** was established in 1970 by the National Academy of Sciences to secure the services of eminent members of appropriate professions in the examination of policy matters pertaining to the health of the public. The Institute acts under the responsibility given to the National Academy of Sciences by its congressional charter to be an adviser to the federal government and, upon its own initiative, to identify issues of medical care, research, and education. Dr. Harvey V. Fineberg is president of the Institute of Medicine.

The **National Research Council** was organized by the National Academy of Sciences in 1916 to associate the broad community of science and technology with the Academy's purposes of furthering knowledge and advising the federal government. Functioning in accordance with general policies determined by the Academy, the Council has become the principal operating agency of both the National Academy of Sciences and the National Academy of Engineering in providing services to the government, the public, and the scientific and engineering communities. The Council is administered jointly by both Academies and the Institute of Medicine. Dr. Ralph J. Cicerone and Dr. Wm. A. Wulf are chair and vice chair, respectively, of the National Research Council.

www.national-academies.org

COMMITTEE ON REVIEW OF NASA'S BIOASTRONAUTICS ROADMAP

DAVID E. LONGNECKER (*Chair*), Association of American Medical Colleges, Washington, DC
JAMES P. BAGIAN, Veterans Health Administration, Ann Arbor, MI
ELIZABETH R. CANTWELL, Los Alamos National Laboratory, NM
VALERIE GAWRON, General Dynamics, Buffalo, NY
CHRISTOPHER A. HART, Federal Aviation Administration, Washington, DC
CHARLES E. LAND, National Cancer Institute, Rockville, MD
DANIEL R. MASYS, Vanderbilt University Medical Center, Nashville, TN
BRUCE M. MCCANDLESS II, Lockheed Martin Space Systems Company, Littleton, CO
TOM S. NEUMAN, University of California, San Diego
THOMAS F. OLTMANNS, Washington University, St. Louis, MO
LAWRENCE A. PALINKAS, University of California, San Diego
JAMES PAWELCZYK, Pennsylvania State University, University Park
BRUCE S. RABIN, University of Pittsburgh, PA
KARLENE ROBERTS, University of California, Berkeley
CAROL E. H. SCOTT-CONNER, University of Iowa Hospitals & Clinics, Iowa City
RHEA SEDDON, Vanderbilt Medical Group, Nashville, TN
JAY R. SHAPIRO, Kennedy-Krieger Institute, Baltimore, MD
THOMAS TEN HAVE, University of Pennsylvania School of Medicine, Philadelphia

IOM Staff

RICARDO A. MOLINS, Study Director (from April 2005)
LISA M. VANDEMARK, Senior Program Officer (through May 2005)
MELVIN H. WORTH, Scholar-in-Residence
BENJAMIN HAMLIN, Research Associate (through April 2004)
ERIN MCCARVILLE, Senior Project Assistant (through June 2005)
VILIJA TEEL, Senior Project Assistant (from July 2005)

COMMITTEE ON AEROSPACE MEDICINE AND THE MEDICINE OF EXTREME ENVIRONMENTS

DAVID E. LONGNECKER *(Chair),* Association of American Medical Colleges, Washington, DC
ALFRED F. CONNORS, JR., Case Western Reserve University, Cleveland, OH
DANIEL R. MASYS, Vanderbilt University Medical Center, Nashville, TN
VAN C. MOW, Columbia University, New York, NY
TOM S. NEUMAN, University of California, San Diego
THOMAS F. OLTMANNS, Washington University, St. Louis, MO
RUSSELL B. RAYMAN, Aerospace Medicine Association, Alexandria, VA
CAROL E. H. SCOTT-CONNER, University of Iowa Hospitals & Clinics, Iowa City
RHEA SEDDON, Vanderbilt Medical Group, Nashville, TN

IOM Staff

CATHARYN LIVERMAN, Study Director
MELVIN H. WORTH, Scholar-in-Residence
NORA HENNESSY, Research Associate
JUDY ESTEP, Senior Project Assistant
VILIJA TEEL, Senior Project Assistant

BOARD ON HEALTH SCIENCES POLICY

PHILIP PIZZO (*Chair*), Stanford University School of Medicine, CA
LESLIE BENET, University of California, San Francisco
DAVID BLUMENTHAL, Harvard Medical School and Massachusetts General Hospital, Boston
GAIL H. CASSELL, Eli Lilly and Company, Indianapolis, IN
ELLEN WRIGHT CLAYTON, Vanderbilt University Medical School, Nashville, TN
DAVID COX, Perlegen Sciences, Mountain View, CA
NANCY DUBLER, Montefiore Medical Center and The Albert Einstein College of Medicine, Bronx, NY
ROBERT GIBBONS, University of Illinois at Chicago
LYNN R. GOLDMAN, Johns Hopkins Bloomberg School of Public Health, Baltimore, MD
BERNARD GOLDSTEIN, University of Pittsburgh, PA
MARTHA N. HILL, Johns Hopkins University School of Nursing, Baltimore, MD
DANIEL R. MASYS, Vanderbilt University Medical Center, Nashville, TN
JONATHAN MORENO, University of Virginia, Charlottesville
E. ALBERT REECE, University of Arkansas, Little Rock
MYRL WEINBERG, National Health Council, Washington, DC
MICHAEL J. WELCH, Washington University School of Medicine, St. Louis, MO
MARY WOOLLEY, Research!America, Alexandria, VA

IOM Staff

ANDREW M. POPE, Director
DAVID CODREA, Financial Associate
AMY HAAS, Administrative Assistant

IOM Boards do not review or approve individual reports and are not asked to endorse conclusions and recommendations. The responsibility for the content of the report rests with the authoring committee and the institution.

SPACE STUDIES BOARD

RICHARD H. TRULY,[†] National Renewable Energy Laboratory, Golden, CO (ret.)
J. CRAIG WHEELER, University of Texas, Austin
A. THOMAS YOUNG, Lockheed Martin Corporation, Onancock, VA (ret.)
GARY P. ZANK,[†] University of California at Riverside

NRC Staff

JOSEPH K. ALEXANDER, Director
SANDRA GRAHAM, Senior Program Officer

---

*Member until June 30, 2005.
†Member starting July 1, 2005.

AERONAUTICS AND SPACE ENGINEERING BOARD

ROBERT E. WHITEHEAD, National Institute of Aerospace, Henrico, NC
THOMAS L. WILLIAMS, Northrop Grumman, El Segundo, CA

NRC Staff

GEORGE LEVIN, Director
MAUREEN MELLODY, Program Officer

Independent Report Reviewers

This report has been reviewed in draft form by individuals chosen for their diverse perspectives and technical expertise, in accordance with procedures approved by the National Research Council's Report Review Committee. The purpose of this independent review is to provide candid and critical comments that will assist the institution in making its published report as sound as possible and to ensure that the report meets institutional standards for objectivity, evidence, and responsiveness to the study charge. The review comments and draft manuscript remain confidential to protect the integrity of the deliberative process. We wish to thank the following individuals for their review of this report:

David W. Gaylor, Gaylor & Associates
Al Harrison, University of California, Davis
John Kindinger, Los Alamos National Laboratory
David M. Klaus, University of Colorado
Mary Jane Osborn, University of Connecticut Health Center
Russell Rayman, Aerospace Medicine Association

Although the reviewers listed above have provided many constructive comments and suggestions, they were not asked to endorse the conclusions or recommendations nor did they see the final draft of the report before its release. The review of this report was overseen by **John R. Ball,** American

Society for Clinical Pathology, and **Richard L. Summers,** University of Mississippi Medical Center. Appointed by the National Research Council and the Institute of Medicine, they were responsible for making certain that an independent examination of this report was carried out in accordance with institutional procedures and that all review comments were carefully considered. Responsibility for the final content of this report rests entirely with the authoring committee and the institution.

Contents

EXECUTIVE SUMMARY 1

1 INTRODUCTION 17
The President's Initiative, 20
Charge to the Committee, 21
Methodology, 22
Overall Assessment of the Bioastronautics Roadmap, 23
Structure of the Report, 24
References, 24

2 CONSIDERATIONS REGARDING THE BR CONTENT 25
Overarching Issues, 26
Fusing the Relationships Between Human Factors and Technology in the BR, 34
Specific Issues, 40
References, 54

3 CONSIDERATIONS REGARDING THE BR PROCESS 60
Risk Assessment, 60
Determination of Acceptable Risk, 67
Risk Communication, 68
Keeping the BR Current, 69
Independent Health and Medical Authority, 70
References, 71

4 CONSIDERATIONS REGARDING THE BR CONTEXT 73
Organizational Characteristics and Risk, 73
Analysis and Prioritization to Meet the Launch Schedule, 75
Addressing the Challenges Posed by the Small Sample Size, 78
Efficiency and Technology Issues, 80
The Case of Advanced Human Life Support, 82
References, 83

APPENDIXES

A Bibliography of Related National Academies Reports 87
B Methods 89
C Status of Countermeasure and Technology Readiness Levels 97
D Bone Fracture Risk Associated with Prolonged Exposure to Microgravity 115
E Integration of Data-Based Evidence and Expert Opinion in Decision Making 120
F Biographical Sketches of Committee Members and Staff 124

INDEX 135

List of Tables, Figures, and Boxes

TABLES

1-1 Bioastronautics Roadmap Risk Rating Categories and Priority Definitions, 20

2-1 Countermeasure Readiness Level (CRL) and Technology Readiness Level (TRL), 30

FIGURES

2-1 Countermeasure and Technology Readiness Levels, 31

3-1 Results of the CB/SD/BSMT Consensus Workshop Held May 25-26, 2004, in Houston, Texas, 66

BOXES

ES-1 Summary of Recommendations, 14

1-1 President Bush's Vision for U.S. Space Exploration, 22

Executive Summary

ABSTRACT *The National Aeronautics and Space Administration's (NASA's) Bioastronautics Roadmap (BR) is "the framework used to identify and assess the risks of crew exposure to the hazardous environments of space." The BR was created to facilitate and support the successful accomplishment of the three Design Reference Missions: a one-year mission to the International Space Station, a month-long stay on the lunar surface, and a 30-month round-trip journey to Mars. The contents of the document are the identified risks, the research and technology questions associated with these risks, and the desired outcomes.*

In 2003, NASA asked the National Academies to conduct a comprehensive assessment of the BR and identify the unique challenges in accomplishing its goals. An ad hoc committee examined the content of the BR, the process used in developing and updating it, and the context in which the BR was developed and will be used.

The committee concluded that the current version of the BR is a useful first step, but it will not be adequate to achieve its stated goals unless the recommendations provided by the committee are incorporated into the document and into the thinking and actions of NASA's management. Four of the committee's 16 recommendations were labeled "principal," indicating their crucial role in a risk mitigation plan. These are (1) accelerate countermeasure and technology development; (2) establish a safe radiation exposure level for all relevant risks; (3) incorporate quality-of-evidence measures, represent risk severity separately from the state of the mitigation strategy or countermeasure, and use standard uncertainty analysis techniques to quantify risk uncertainty; and (4) ensure that the BR is a dynamic and current database.

The committee identified both overarching and specific issues in need of attention in the BR content. Overarching issues involve the impact of various time factors on risk, the interactions among risks, and the need to create two new cross-cutting categories of risk: "Human Systems Integration" and "Food and Nutrition." Specific issues include the need to (1) validate current and future crew selection criteria; (2) group behavioral health risks into categories based on clinical outcomes and address issues of human sexuality in long-duration missions; (3) use actuarial data to estimate the likelihood of intrinsic health alterations as part of the selection criteria for the Mars mission crew; and (4) quantitatively evaluate mental and physical health risks affecting crew health and mission success.

The committee concurred with NASA regarding the establishment of an independent health and medical authority to enhance the BR process and recommended that NASA (1) add human performance failure due to organizational and cultural factors as a new risk; (2) conduct periodic assessment of additional risks from lack of resources and use this to make decisions about research support; (3) use Bayesian sequential trials approach and hierarchical random or fixed effect methods to address the small sample size resulting from limited opportunities for space flight; and (4) reframe risks as either health or technology related, in order to address issues within the BR context.

INTRODUCTION

Extending the spatial and temporal boundaries of human space flight are important goals for the nation and for the National Aeronautics and Space Administration (NASA). However, human space flight remains an endeavor with substantial risks, and these risks must be identified, managed, and mitigated appropriately to achieve the nation's goals in space. The Bioastronautics Roadmap (hereafter referred to as the BR) is described by NASA as "the framework used to identify and assess the risks of crew exposure to the hazardous environments of space" (NASA, 2005). The BR also "guides the prioritized research and technology development that, coupled with operational space medicine, will inform the development of medical standards and policies, the specifications of requirements for the human system, and the implementation of medical operations."

The BR was created to facilitate and support the successful accomplishment of the three Design Reference Missions described in the President's space initiative of January 14, 2004 (White House, 2004), specifically, a one-year mission to the International Space Station (ISS), a month-long stay on the lunar surface, and a 30-month round-trip journey to Mars. The stated goal of the BR is "to reduce risk through effective and efficient mitigation solutions developed from a focused research and technology development strategy" (NASA, 2005). The contents of the document are the identified risks; the research and technology questions associated with these risks; and the deliverables—or the desired outcomes or solutions—related to these questions. Major processes of the BR include risk identification and risk assessment. The context in which these risks are identified includes the mission requirements for the three Design Reference Missions as well as the organization and systems within NASA and the external organizations and systems that govern NASA (e.g., executive and legislative branches of government, federal budget).

The February 2005 version of the BR used by the committee for its review, in a traditional document format, is enclosed as a CD in the cover of this report. Both the baseline document and the interactive version that relates to specific risks and Design Reference Missions are available on-line at *http://bioastroroadmap.nasa.gov.*

CHARGE TO THE COMMITTEE

In 2003, prior to the unveiling of the President's space exploration vision, NASA asked the Institute of Medicine (IOM), in collaboration with the Division on Engineering and Physical Sciences of the National Academies, to conduct a review of the BR. Specifically, NASA asked the committee to (1) conduct a comprehensive assessment and report of the strengths and weaknesses of the content and processes of the BR as applied to the missions described in the President's exploration initiative and (2) identify the unique challenges for accomplishing its goals and objectives. Specific questions for the committee to answer included—but were not limited to—the following:

1. How can the BR better capture and describe the critical risks and key research and technology issues for risk reduction and management so as to provide a framework for informed decisions regarding resource allocation?

2. Does the BR use an appropriate method of risk assessment and expression of risk assessment? Does it adequately communicate the methods underlying risk assessment and the resulting activities for different mission scenarios?

3. How well does the BR address different types of risk (e.g., health, engineering) and their impact?

4. Are the categories of critical research issues and the metrics used to analyze them appropriate (risk assessment and characterization, mechanistic/process research, countermeasure development, and medical diagnosis and treatment)?

5. Are efficiency and technology issues properly and adequately addressed?

Responding to NASA's request, the Committee on Review of NASA's Bioastronautics Roadmap approached its charge by focusing on the three general categories of issues in the content of the BR (Chapter 2); the process involved in the development, updating, and utilization of the BR (Chapter 3); and the context within which the BR is framed and expected to be used (Chapter 4). The committee held eight meetings, most of which included a data-gathering session where testimony from NASA officials and space science experts was heard. In addition, several committee members met with NASA officials and other experts to gather information.

The committee's conclusions and recommendations in this report are based on published sources of information as well as on the committee's experience and expertise in the diverse fields included in the BR.

OVERALL ASSESSMENT OF THE BIOASTRONAUTICS ROADMAP

The committee's comments focus on areas in which improvement seems necessary and most valuable, but the committee wishes to emphasize that the present document demonstrates that NASA has considered many of the key factors carefully and is handling many aspects of this complex challenge appropriately. The committee was impressed by the progress that appeared in the BR during the course of this review. The reader should not lose sight of these many positives when reading the following analyses, conclusions, and recommendations, which focus on opportunities for improvement. The current version of the BR is a useful first step, but it will not be adequate to achieve its stated goals unless the recommendations provided here are incorporated into the document and into the thinking and actions of NASA management. The BR must constantly be updated and maintained, the resulting action plans that flow from the BR must be supported by adequate allocation of resources both to NASA and within NASA, and the action plans must be implemented fully. If these criteria are met, the committee believes that the BR will be an effective mechanism to mitigate risks to human health and thus contribute to ensure mission success during extended space flight.

PRINCIPAL FINDINGS AND RECOMMENDATIONS

The committee concludes that all of its recommendations should be implemented in order to achieve the goals of the BR. However, the committee is convinced that the following four principal recommendations must be applied if the BR is to be an effective approach to risk identification and risk reduction associated with the Design Reference Missions.

Status of Readiness Levels

A variety of deliverable products result from the BR. Progress in these areas is gauged by establishing "readiness levels" that delineate the level of maturity of countermeasures or technologies to support the human system in space. Currently, the BR proposes 183 projected deliverables. In the context of the BR, these deliverables constitute the risk mitigation plan for the human system. The BR emphasizes that "roadmap activities must focus on operational issues and solutions to operational problems to support an outcome-oriented approach." Thus, "bioastronautics research is focusing on deliverables at a readiness level of 4 or greater" (NASA, 2005, p. 17). However, more than half (53%) of the deliverables proposed in the BR rank below the stage 4 level of readiness and thus below the threshold for priority in bioastronautics research consideration. The committee concludes that this emphasis on an applied research agenda for NASA bioastronautics is not without significant consequences and risks, especially given the relatively immature status of current countermeasure development. Further, the committee finds that the majority of projected BR countermeasures, mitigations, or other deliverables are in a nascent state of readiness (i.e., below the threshold for prioritization in bioastronautics research) and that the state of countermeasure development significantly lags the need. Current resources are unlikely to be sufficient to complete the BR mitigation plan in a time frame that enables the exploration class missions envisioned by NASA.

> *Recommendation 2.3*—Accelerate Countermeasure and Technology Development
> **The committee recommends that NASA initiate an aggressive program, including the use of animal models, analog environments, and space flight, to significantly accelerate the progress of all Countermeasure Readiness Levels and Technology Readiness Levels that are essential to support the proposed exploration agenda. Countermeasures and technologies at an undefined or low state of readiness (the majority of the current portfolio) should receive renewed attention. The committee notes further that failure to do so will jeopardize the exploration program outlined in the President's vision for exploration of January 2004.**

Radiation Effects—
Establishing Risk-Specific Radiation Exposure Levels

The committee concludes that radiation effects are given ample weight in the BR and concurs with NASA's assessment that the duration of the mission and the distance from Earth will determine the amount and type of radiation exposure and the required shielding for both the vehicle and the protective wear for extravehicular activity (EVA), with the associated weight and design implications. However, the conventional rule of thumb for terrestrial radiation protection—that is, that protection against late radiation effects such as increased cancer risk will also protect against acute radiation effects—may not hold for high-energy radiation (HZE) from galactic cosmic rays, particularly regarding central nervous system (CNS) impairment. As with the other components of the BR, it will be essential to follow the developments in both space and terrestrial biological research as well as shielding technology to ensure crew health and safety for longer-duration, higher-radiation-exposure exploratory missions.

Recommendation 2.9—Establish a Safe Radiation Exposure Level for All Relevant Risks

The committee recommends that a safe radiation exposure level be established by NASA for each relevant risk, based on projected flight duration and distance from Earth, and that the technology to keep the level of exposure below that limit be ensured. Inherent in this recommendation and consistent with Recommendation 2.3, the committee concludes that NASA must conduct further research to clarify the extent to which protracted or low-dose HZE radiation exposure might contribute to mission-damaging CNS effects.

Risk Assessment

The committee agrees that NASA's decision to draw on expert opinion in identifying and ranking risks is a reasonable strategy. However, the committee believes that there are weaknesses in the current risk assessment process related to (1) lack of information regarding the quality of evidence that informs risk assessment, (2) the obscuring of risk that results from "lumping" both the risk and its associated mitigation into a single value, and (3) lack of a quantitative measurement of uncertainty related to the risk. These areas can, and should, be enhanced. Risk assessment should primarily be evidence-based wherever possible, and where evidence does not exist, research should be directed to acquiring the evidence needed. Rating the quality of each published source of information should be a component of the BR.

The committee also believes that disaggregation of risk from mitigation of risk is important because (1) aggregated risk values obscure the risk itself and can lead to false confidence or concern, depending on the status of countermeasure or technology development; (2) approaches that mitigate one risk may have a positive or negative impact on other risks or mitigations; and (3) changes in systems may have a positive or negative impact on mitigations. Failure to track risk separately from risk mitigation could well lead to failure to focus on the inherent relationships among risks and mitigations.

In addition, the determination of risk always involves an element of uncertainty. The committee concludes that the current printed and on-line versions of the BR do not include any expression of uncertainty in terms of risk estimates, reported confidence intervals, or narrative discussion.

Recommendations 3.1, 3.2, and 3.3—Incorporate Quality-of-Evidence Measures, Represent Risk Severity Separately from the State of the Mitigation Strategy or Countermeasure, and Use Standard Uncertainty Analysis Techniques to Quantify Risk Uncertainty

The committee recommends that NASA (1) determine, and incorporate into the BR, measures of the quality of the evidence that form the basis for defining risks and the assessments associated with each risk; (2) structure the BR to represent separately the severity and likelihood of each risk and the state of the mitigation strategy or countermeasures associated with each risk; and (3) whenever possible, restructure the BR to include a quantification of the uncertainty of risks using standard uncertainty analysis techniques (e.g., frequentist, Bayesian, or possibility theory and approximate reasoning) that will provide uncertainty distributions or ranges in addition to point estimates. This will help contribute to the subsequent definition of operating bands.

Risk Communication and Keeping the BR Current

The way in which risk-related information is represented in the BR and communicated to users is important to its overall effectiveness as a program management tool. One widely used format for representing risks not currently incorporated into the BR is the NASA-wide Continuous Risk Management Program, and the committee encourages its continued use because it is widely recognized and understood throughout NASA. The committee believes that the BR is better thought of and designed as a dynamic database of information relative to risk definition and assessment, from which a document or set of alternative documents can be derived at any time and incorporated into a risk management program. The web-based on-line version of the BR is an important step in this direction.

It is fundamentally important that configuration control methods be established and implemented to keep the BR up-to-date as new knowledge and technologies develop. This process can be facilitated by identifying an "owner and manager" within NASA for each set of related BR risks and establishing a regular review cycle that should occur not less than once a year. Where there is a desire to combine published research data with "expert opinion" from stakeholders, methods such as Bayesian updating and elicitation of expert opinion are available.

Recommendations 3.4 and 3.5—Ensure That the BR Is a Dynamic and Current Database That Enhances Communication and Conveys Information Effectively
To enhance effective communication of the content of the BR, the committee recommends that the BR be designed and utilized as a dynamic database of information relative to risk definition and assessment, from which a document or set of alternative documents can be derived at any time and incorporated into a risk management program.

The references currently cited in the BR must be updated to reflect more recent research on both risk identification and countermeasure development. Moreover, a mechanism should be established for the *ongoing review of the current best evidence* contained in the research literature, and methods should be developed to integrate new findings from the literature with the expert opinion of key stakeholders, including those from operations and the research community.

The remaining recommendations focus on either general principles (overarching issues) or specific issues that need to be addressed in the BR. The committee believes that all of these are equally important and should be implemented; thus, they are not presented in order of priority but rather are grouped by the three principal themes of content, process, and context.

OVERARCHING ISSUES IN THE BR CONTENT

Overarching issues are those factors that, in the committee's view, deserve wide review and application throughout the current BR and future revisions of the BR. They should be viewed as strategic approaches to the revision and management of the BR.

The Time Factor and Its Impact on Risk

Recommendation 2.1—Label Risks by Relevance to Operational Requirements and Temporal Urgency
The committee recommends that risk assessment and mitigation (technology or countermeasure development) in the BR be enhanced by labeling risks according to their relevance to operational requirements and to temporal urgency.

Interactions Among Risks

Recommendation 2.2—Define, Quantify, and Mitigate Interactions Among Risks
The committee recommends that greater effort be devoted to identifying and explaining the interrelations among risks and risk mitigations that are grouped within and across the cross-cutting categories in the BR.

Creating the Cross-Cutting Categories "Human Systems Integration" and "Food and Nutrition"

Recommendations 2.4 and 2.5—Create Two New Cross-Cutting Categories: Human Systems Integration and Food and Nutrition
The committee recommends that NASA create two new cross-cutting categories in the BR:

1. A cross-cutting category that spans the two existing categories of "Behavioral Health and Performance" and "Space Human Factors Engineering" risks listed currently in the area of Advanced Human Support Technologies. This new category should be labeled "Human Systems Integration," consistent with the terminology currently in use by the U.S. Department of Defense (2004) and the Human Systems Integration in NASA Headquarters.
2. The cross-cutting area "Nutrition" should be renamed and expanded to "Food and Nutrition" to more fully encompass the concepts of dietary needs for space flight and emphasize the relationships between nutrition and food technology. Further, the committee recommends that the impact of inadequate food and nutrition on processes related to mental and physical health risks and maintenance of the space environment and life support systems be defined from the standpoints of food safety, quality, and quantity, and interventions proposed if problems occur.

SPECIFIC ISSUES

Specific issues are those areas that deserve focused attention and refinement in the BR to more effectively accomplish its goals as a management tool. They should be viewed as tactics for the document, whereas the overall issues should be viewed as strategies.

Recommendations Related to Specific Issues in the BR Content

Recommendation 2.6—Validate Current and Future Crew Selection Criteria

The committee recommends that the Astronaut Office and representative flight surgeons be consulted regarding the crew selection process in order to place greater emphasis on the roles of crew compatibility and team performance in overall mission success.

Recommendations 2.7 and 2.8—Group Behavioral Health Risks into Categories Based on Clinical Outcomes, and Address Issues of Human Sexuality in Long-Duration Missions

The committee recommends that behavioral health risks within the proposed new cross-cutting category of Human Systems Integration be grouped into categories based on clinical outcomes such as interpersonal conflict, affect regulation, decrements in cognitive performance, mood disorders, and sleep disorders, rather than categories such as psychosocial, neurobehavioral, cognitive, and circadian rhythms, and that the interrelations between these categories be delineated clearly. In addition, issues of human sexuality should be addressed in the BR in relation to long-duration missions such as the proposed Mars Design Reference Mission.

Recommendation 2.10—Use Actuarial Data to Estimate the Likelihood of Intrinsic Health Alterations as Part of the Selection Criteria for the Mars Mission Crew

The committee recommends that, wherever possible, NASA use actuarial data to estimate and/or model the likelihood of intrinsic health alterations for crew who will be part of the Mars mission. Utilization of this information as part of the selection criteria for astronauts should be considered. After intrinsic health risks are estimated, NASA should then estimate and/or model the contribution of the space environment and life support system malfunction to increased risk.

Recommendation 2.11—Develop a System for Quantitative Evaluation of Mental and Physical Health Risks Affecting Crew Health and Mission Success

The committee recommends that a system be developed for quantitatively evaluating the mental and physical health risks that could affect mission success and crew health and that priorities for countermeasure development (i.e., definitive treatment vs. palliation) be established for the most likely conditions to be encountered during each reference mission. A panel of outstanding medical clinicians should be used to assist NASA medical operations staff in characterizing the likelihood, importance, and "treatability" of each condition.

Recommendations Related to Specific Issues in the BR Process

Recommendation 3.6—Establish an Independent Health and Medical Authority

The committee endorses the principle that critical decisions regarding safety and health should be made by an authority that is independent of programmatic costs and schedules. Given the importance and complexity of health and human safety issues, the committee acknowledges and endorses the creation of the Independent Health and Medical Authority (IHMA), analogous to the Independent Technical Authority, and recommends that the IHMA be given responsibility, authority, and accountability for the health and safety decisions that relate to risks identified in the BR.

Recommendations Related to Specific Issues in the BR Context

Recommendation 4.1—Add New Risk: Human Performance Failure Due to Organizational and Cultural Factors

The committee recommends that an additional risk labeled "human performance failure due to organizational and cultural factors" be added to the BR. It may prove optimal to track this risk in a manner differently from the other risks in the BR (e.g., annual analyses of organizational and cultural risk in a separate report, use of an external standing panel to discuss this issue regularly). The committee's intent is that a

risk-focused analysis of organizational and cultural issues become a visible part of the BR process.

Recommendation 4.2—Conduct Periodic Assessment of Additional Risks from Lack of Resources and Use This to Make Decisions About Microgravity and Behavioral Research Support
The committee recommends that NASA perform regular, detailed assessments of the additional risks to the conduct of the President's 2004 vision for space exploration posed by the lack of available resources to fully address the issues posed in the BR. This assessment should then be used to make early strategic decisions regarding issues such as, but not limited to, the following:

1. **How to provide support for a microgravity research platform that will have the resources (crew time, up-mass, facilities, and power) for the large amount of work necessary to validate countermeasures; achieve Technology Readiness Level 7 for life support systems sufficiently early in the design phase to allow their integration into the overall vehicle; and demonstrate the utility of medical procedures in microgravity.**
2. **How to support the extensive behavioral research program that would be necessary to validate processes or countermeasures such as select-in–select-out criteria (both for individual crew members and for a composite crew), issues related to cultural diversity, crew interactions, and isolation or stress-induced hazards. These issues may well require long lead times to study adequately.**

Recommendation 4.3—Use Bayesian Sequential Trials Approach and Hierarchical Random or Fixed Effects Methods to Address the Small Sample Size
Drawing on the findings of the Institute of Medicine report *Small Clinical Trials: Issues and Challenges* (IOM, 2001), the committee recommends the use of pooled data from Bayesian sequential trials techniques and hierarchical random or fixed effects methods to compensate for the small sample sizes associated with individual flights.

Recommendation 4.4—Reframe Risks as Either Health or Technology Related

The committee recommends that the current definition of risk be altered to clearly identify at least two types of risks: (1) health and medical risk, defined as the conditional probability of an adverse event to the human system (i.e., crew health or medical event) resulting from exposure to the space flight environment, and (2) engineering technology and system performance risk, defined as the conditional probability of an adverse event resulting from the space flight supersystem that affects crew health or mission success.

Box ES-1 contains a summary of the report's recommendations.

BOX ES-1
SUMMARY OF RECOMMENDATIONS

Principal Recommendations

2.3 Accelerate Countermeasure and Technology Development
2.9 Establish a Safe Radiation Exposure Level for All Relevant Risks
3.1, 3.2, 3.3 Incorporate Quality-of-Evidence Measures, Represent Risk Severity Separately from the State of the Mitigation Strategy or Countermeasure, and Use Standard Uncertainty Analysis Techniques to Quantify Risk Uncertainty
3.4, 3.5 Ensure That the BR Is a Dynamic and Current Database That Enhances Communication and Conveys Information Effectively

Recommendations on Overarching Issues in the BR Content

2.1 Label Risks by Relevance to Operational Requirements and Temporal Urgency
2.2 Define, Quantify, and Mitigate Interactions Among Risks
2.4, 2.5 Create Two New Cross-Cutting Categories: Human Systems Integration and Food and Nutrition

BOX ES-1 Continued

Recommendations on Specific Issues in the BR Content

2.6 Validate Current and Future Crew Selection Criteria
2.7, 2.8 Group Behavioral Health Risks into Categories Based on Clinical Outcomes, and Address Issues of Human Sexuality in Long-Duration Missions
2.10 Use Actuarial Data to Estimate the Likelihood of Intrinsic Health Alterations as Part of the Selection Criteria for the Mars Mission Crew
2.11 Develop a System for Quantitative Evaluation of Mental and Physical Health Risks Affecting Crew Health and Mission Success

Recommendations on Specific Issues in the BR Process

3.6 Establish an Independent Health and Medical Authority

Recommendations on Specific Issues in the BR Context

4.1 Add New Risk: Human Performance Failure Due to Organizational and Cultural Factors
4.2 Conduct Periodic Assessment of Additional Risks from Lack of Resources and Use This to Make Decisions About Microgravity and Behavioral Research Support
4.3 Use Bayesian Sequential Trials Approach and Hierarchical Random or Fixed Effects Methods to Address the Small Sample Size
4.4 Reframe Risks as Either Health or Technology Related

REFERENCES

IOM (Institute of Medicine). 2001. *Small Clinical Trials: Issues and Challenges*. Washington, DC: National Academy Press.

NASA (National Aeronautics and Space Administration). 2005. Bioastronautics Roadmap—a risk reduction strategy for human space exploration. On-line [available: *http://ston.jsc.nasa.gov/collections/TRS/-tecjre[Sp-2005-6113.pdf*]. Accessed 1/6/2006.

U.S. DOD (U.S. Department of Defense). 2004. Defense Acquisition Guidebook. DoDD 5000.1 On-line [available: *http://akss.dau.mil/dag*].

White House. 2004. President Bush announces new vision for space exploration program. Remarks by the President on U.S. space policy. On-line [available: *http://www.whitehouse.gov/news/releases/2004/01/20040114-3.html*]. Accessed 5/26/05.

1

Introduction

Extending the spatial and temporal boundaries of human space flight is an important goal for the nation and for the National Aeronautics and Space Administration (NASA). However, human space flight remains an endeavor with substantial risks, and these risks must be identified, managed, and mitigated appropriately to achieve the nation's goals in space. The Bioastronautics Roadmap (BR) is the result of extensive, commendable efforts on the part of NASA to prioritize research efforts to meet these challenges. It is a broad and complex document that has been developed with care and thought and has evolved over time as the thinking at NASA has progressed regarding its role. During the time this committee was active, the Bioastronautics Critical Path Roadmap of April 2004 (NASA, 2004) evolved into the Bioastronautics Roadmap of February 9, 2005 (NASA, 2005). The former contained 50 risks, 5 cross-cutting areas, and 1,414 "enabling questions" for the three Design Reference Missions: the International Space Station (ISS), the lunar mission, and Mars. The latter contains 45 risks, the same 5 cross-cutting areas, and 1,360 research and technology questions. Other changes occurred as well, as NASA reviewed and revised its own efforts, and in response to the committee's interim report of 2005 (IOM, 2005). For example, the concepts of "requirements" and "operating bands" were strengthened considerably as the BR evolved. Thus, the committee faced a considerable challenge in responding to its charge to review the BR and its related thought processes because these were dynamic and evolving during most of the review process. However, the committee recognizes that the ongoing changes represent progress that is both necessary and appropriate, and it acknowledges the continued focus on this approach as a management tool for bioastronautics. The committee's comments are, therefore, based on the Bioastronautics Roadmap version of February 9, 2005, the current version at the time of this writing (NASA, 2005), referred to as the BR. The February 2005 baseline version of the BR used by the committee for its review is enclosed as a CD in the cover of this report. Both the baseline BR and the interactive version that relates to specific risks and Design Reference Missions are available on-line at *http://bioastroroadmap.nasa.gov*.

The Bioastronautics Roadmap was developed collaboratively by NASA's Office of Biological and Physical Research and the Office of Space Flight with the concurrence of the Office of the Chief Health and Medical Officer. NASA describes the document as "the framework used to identify and assess the risks of crew exposure to the hazardous environments of space" (NASA, 2005). According to the baseline BR, the BR also "guides the prioritized research and technology development that, coupled with operational space medicine, will inform" the following:

1. The development of medical standards and policies
2. The specifications of requirements for the human system
3. The implementation of medical operations

The BR also provides information that helps establish operating bands—or exposure limits—for humans exposed to risks during space travel and develop countermeasures to maintain the health and functioning of the crew within those limits and technologies to improve the safety and productivity of human space flight. Operating bands represent an acceptable range of performance or functioning, specifically for life support and habitation systems. Exposure limits describe an acceptable maximum decrement or change in a human physiological or behavioral parameter that impacts the performance of assigned tasks or has implications for lifetime medical status.

The BR was created to facilitate and support the successful accomplishment of the three Design Reference Missions described in the President's space initiative of January 14, 2004 (White House, 2004). The stated goal of the BR is "to reduce risk through effective and efficient mitigation solutions developed from a focused research and technology development strategy" (NASA, 2005). The contents of the BR are the identified risks; the research and technology questions associated with these risks; and the deliverables—the desired outcomes or solutions to these questions. Major processes of the BR include risk identification and risk assessment. The context in which these risks are identified include the mission requirements for the three Design Reference Missions as well as the organization and systems within NASA and the external organizations and systems that govern NASA (e.g., executive and legislative branches of government, federal budget).

Efforts to understand and manage the risks associated with human space flight have been ongoing at NASA for many years, and specific activities related to the development of a roadmap began in the early 1990s. The process of risk identification that resulted in the BR commenced in 1997 in brainstorming sessions involving NASA and non-NASA experts who rated risks within their own discipline areas. With guidance from NASA and other advisory reports (see Appendix A), 150 risks were identified. More recently, and after several iterations, the list was culled to the 45 risks that are the focus of the current BR. The BR currently identifies 31 human health–related risks and 14 risks related to systems performance and efficiency clustered in 5 cross-cutting areas: human health and countermeasures, radiation health, behavioral health and performance, autonomous medical care, and advanced human life support technologies.

The final risks and related research questions were identified by discipline-specific teams using internal NASA and external advisory committee reports, as well as other recent research findings. The Bioastronautics Science Management Team, which includes NASA scientists, managers, and flight surgeons, and the National Space Biomedical Research Institute (NSBRI) director reviewed and discussed the risks and provided oversight for the project. In the spring of 2004, NASA held several consensus workshops, which included the research community and NASA operations communities (flight surgeons, astronauts, and the medical office), to address the sample size needed for research related to the risks identified, the use of animal models, and the ranking of biomedical risks from the point of view of astronauts and flight surgeons. In addition, NASA sought comments on the BR from the relevant research communities in a web-based query.

Risk assessment and rating are described in the BR for two general types of risk: human health risks and system performance/efficiency risks. The ratings for the human health risks derive from the analysis of the likelihood of occurrence of each risk, the severity of the consequences should a

TABLE 1-1 Bioastronautics Roadmap Risk Rating Categories and Priority Definitions

Risk Rating Priority	Human Health Risks	System Performance or Efficiency Risks
1	Risk of serious adverse health or performance consequences, and there is no mitigation strategy that has been validated in space or demonstrated on Earth.	Considerable potential for improvement in mitigation efficiency in many areas; proposed missions may be infeasible without improvements.
2	Risk of serious adverse health or performance consequences, and there is no mitigation strategy that has been validated in space.	Considerable potential for improvement in mitigation efficiency in a few areas.
3	Health and performance consequences are known or suspected but will not affect mission success due to effective mitigation strategies that have been validated in space.	Minimum potential or limited need for improvement in mitigation efficiency.

SOURCE: NASA (2005, Table 7-2).

given event occur, and the status of efforts to mitigate each risk. For the systems risks, the criterion is improved efficiency. Input from the iterative process described above, including results of the workshops, fed into this risk assessment. Table 1-1 shows the risk rating categories and priority definitions used in the BR.

The intent of this risk rating process is to aid communication and decision making by demonstrating the consensus on the relative importance of each risk. The same criteria were applied to all 45 risks in the current BR.

THE PRESIDENT'S INITIATIVE

On January 14, 2004, President George W. Bush announced his vision for space exploration (White House, 2004). The President's plan for continued human and robotic space exploration is summarized in Box 1-1. The BR refers to three scenarios in the plan as "Design Reference Missions" and describes them as follows: (1) a one-year mission to the International Space Station (ISS); (2) a one-month stay on the lunar surface; and (3) a 30-month journey to Mars and back.

CHARGE TO THE COMMITTEE

In 2003, NASA asked the Institute of Medicine (IOM), in collaboration with the Division on Engineering and Physical Sciences of the National Academies, to conduct a review of the BR. Specifically, NASA asked the committee to (1) conduct a comprehensive assessment and report of the strengths and weaknesses of the content and processes of the Bioastronautics Roadmap as applied to the missions described in the President's exploration initiative and (2) identify the unique challenges for accomplishing its goals and objectives. Specific questions for the committee to answer included—but were not limited to—the following:

1. How can the BR better capture and describe the critical risks and key research and technology issues for risk reduction and management so as to provide a framework for informed decisions regarding resource allocation?
2. Does the BR use an appropriate method of risk assessment and expression of risk assessment? Does it adequately communicate the methods underlying risk assessment and the resulting activities for different mission scenarios?
3. How well does the BR address different types of risk (e.g., health, engineering) and their impact?
4. Are the categories of critical research issues and the metrics used to analyze them appropriate (risk assessment and characterization, mechanistic/process research, countermeasure development, and medical diagnosis and treatment)?
5. Are efficiency and technology issues properly and adequately addressed?

In September 2004, the committee released its preliminary report to NASA entitled *Preliminary Considerations Regarding NASA's Bioastronautics Critical Path Roadmap* (IOM, 2005). That document presented the committee's preliminary conclusions about the strengths and weaknesses of

BOX 1-1
President Bush's Vision for U.S. Space Exploration

The President's plan for steady human and robotic space exploration is based on the following goals:

1. First, America will complete its work on the International Space Station by 2010, fulfilling our commitment to our 15 partner countries. The United States will launch a re-focused research effort on board the International Space Station to better understand and overcome the effects of human space flight on astronaut health, increasing the safety of future space missions. To accomplish this goal, NASA will return the Space Shuttle to flight consistent with safety concerns and the recommendations of the Columbia Accident Investigation Board. The Shuttle's chief purpose over the next several years will be to help finish assembly of the Station, and the Shuttle will be retired by the end of this decade after nearly 30 years of service.

2. Second, the United States will begin developing a new manned exploration vehicle to explore beyond our orbit to other worlds—the first of its kind since the Apollo Command Module. The new spacecraft, the Crew Exploration Vehicle, will be developed and tested by 2008 and will conduct its first manned mission no later than 2014. The Crew Exploration Vehicle will also be capable

the April 2004 version of the BR. The present report builds on those preliminary conclusions and provides recommendations to NASA about how to address the issues identified by the committee. The present report refers to the February 9, 2005, version of the BR.

METHODOLOGY

Responding to NASA's request, the committee approached its task with enthusiasm and a strong sense of commitment to the goals of NASA overall and to its most visible images, the astronauts. The efforts of the committee were buoyed by the excitement associated with the President's exploration initiative of January 2004. The committee included members with a broad range of relevant expertise, and it supplemented that expertise by holding eight meetings in either open session (see Appendix B) or executive session, of transporting astronauts and scientists to the International Space Station after the Shuttle is retired.

3. Third, America will return to the Moon as early as 2015 and no later than 2020 and use it as a stepping stone for more ambitious missions. A series of robotic missions to the Moon, similar to the Spirit Rover that is sending remarkable images back to Earth from Mars, will explore the lunar surface beginning no later than 2008 to research and prepare for future human exploration. Using the Crew Exploration Vehicle, humans will conduct extended lunar missions as early as 2015, with the goal of living and working there for increasingly extended periods. The extended human presence on the Moon will enable astronauts to develop new technologies and harness the Moon's abundant resources to allow manned exploration of more challenging environments. An extended human presence on the Moon could reduce the costs of further exploration, since lunar-based spacecraft could escape the Moon's lower gravity using less energy at less cost than Earth-based vehicles. The experience and knowledge gained on the Moon will serve as a foundation for human missions beyond the Moon, beginning with Mars. NASA will increase the use of robotic exploration to maximize our understanding of the solar system and pave the way for more ambitious manned missions. Probes, landers, and similar unmanned vehicles will serve as trailblazers and send vast amounts of knowledge back to scientists on Earth.

where experts from NASA and elsewhere presented their views on relevant content of the BR; by individual meetings between subgroups of the committee, NASA personnel, and outside experts; by more than 60 hours of deliberations in executive session; and by countless hours of independent reading, analysis, and team writing. This report represents the synthesis of these efforts and the collective agreement of committee members.

OVERALL ASSESSMENT OF THE BIOASTRONAUTICS ROADMAP

The committee's comments focus on areas in which improvement seems necessary and most valuable, but the committee wishes to emphasize that the present document demonstrates that NASA has considered many of the key factors carefully and is handling many aspects of this complex challenge appropriately. The committee was impressed by the progress that appeared in the BR during the course of its review. The reader should not lose sight of these many positives when reading the following analyses, conclusions, and recommendations, which focus on opportunities for improvement. The current version of the BR is a useful first step, but it will not be adequate to achieve its stated goals unless the recommendations provided here are incorporated into the document and into the thinking and actions of NASA management. The BR must constantly be updated and maintained, the resulting action plans that flow from the BR must be supported by adequate allocation of resources both to NASA and within NASA, and the action plans must be implemented fully. If these criteria are met, the committee believes that the BR will be an effective mechanism to mitigate the risks to human health and thus contribute to ensuring mission success during extended space flight.

STRUCTURE OF THE REPORT

The report has been structured according to the committee's charge insofar as the BR content (Chapter 2) and process (Chapter 3) are concerned. Chapter 4 deals with issues relevant to the BR context and thus addresses what the committee views as the unique challenges faced by NASA in accomplishing the roadmap's goals and objectives. Specific questions posed in the charge are addressed in the relevant sections of this report. The Summary presents the committee's finding and recommendations, highlighted throughout the report.

REFERENCES

IOM (Institute of Medicine). 2005. *Preliminary Considerations Regarding NASA's Bioastronautics Critical Path Roadmap*. . Washington, DC: The National Academies Press.

NASA (National Aeronautics and Space Administration). 2005. Bioastronautics Roadmap—a risk reduction strategy for human space exploration. On-line [available: *http://ston.jsc.nasa.gov/collections/TRS/-techrep/Sp-2005-6113.pdf*]. Accessed 1/6/2006.

White House. 2004. President Bush announces new vision for space exploration program. Remarks by the President on U.S. space policy. On-line [available: *http://www.whitehouse.gov/news/releases/2004/01/20040114-3.html*]. Accessed 5/26/05.

2

Considerations Regarding the BR Content

The major content areas of the current Bioastronautics Roadmap (BR) include 45 risks that are assigned to one of 5 cross-cutting areas: (1) Human Health and Countermeasures, (2) Autonomous Medical Care, (3) Behavioral Health and Performance, (4) Radiation Health, and (5) Advanced Human Support Technologies (NASA, 2005). Although not emphasized in the body of the BR, the associated appendixes provide data regarding technology readiness levels and countermeasure readiness levels (NASA, 2005, Appendix A) associated with each risk; indicate some anticipated interactions among risks (NASA, 2005, Appendix B); and depict a preliminary schedule of deliverables for the 5 cross-cutting areas (NASA, 2005, Appendix C). The risks are further described by their associated research and technology questions, and all of the above are further analyzed relative to the Design Reference Missions. Together, these areas result in a multidimensional matrix that challenges both National Aeronautics and Space Administration (NASA) managers and the review committee, and the challenge is further compounded by the dynamic nature of the present and future versions of the BR.

The committee's comments for content improvement are divided into two broad categories: overarching issues and specific issues. The overarching issues involve (1) the time dimensions of risk, (2) the interactions among risks, (3) the status of the countermeasure and technology readiness levels, and (4) linking relationships between human factors and technology in the BR. Attention to each of these overarching issues will strengthen the content of the BR and provide a framework for thinking that will benefit NASA strategists, managers, and operations personnel as they focus on risk reduction related to the ambitious objectives outlined in the President's initiative of January 2004 (White House, 2004). Specific areas in which additional attention would be of great benefit include the following:

- Reclassification of behavioral health risks
- Psychological and physical impacts of space flight on performance, including use of crew selection criteria (social, demographic, and preexisting health status of astronauts, and their response to stress) to minimize adverse responses, especially in the context of longer-term missions
- Radiation effects—establishing risk-specific radiation exposure levels
- Assessing the sources and impact of long-duration space flight on crew health and incremental risk
- Autonomous medical care and self-care

OVERARCHING ISSUES

Overarching issues are those factors that, in the committee's view, deserve wide review and application throughout the current BR and in future revisions of the BR. They should be viewed as guiding principles or strategic approaches to the revision and management of the BR.

The Time Factor and Its Impact on Risk

Time is a factor that has the potential to increase risk significantly, particularly in the context of long-duration space flight such as the 30-month Mars mission outlined in the President's initiative. Time has several dimensions that must be considered in the definition and mitigation of risk. Duration of the mission is one component. Clearly, the potential for the development of a health problem, such as new disease or injury, increases progressively from the 1-month lunar mission to the 12-month International Space Station (ISS) mission, to the 30-month Mars mission. Similarly, the consequences and countermeasures associated with disease or injury will differ depending on the time of appearance of the human health problem within the mission time frame (consider, for example, the discovery of a breast mass that appears 2 months after launch on a planned 30-

month Mars mission, versus one that occurs 2 weeks into a planned 30-day lunar mission).

Another dimension of the time factor relates to distance from Earth during the mission. Distance from Earth impacts the mitigation strategy and selection of appropriate countermeasures because increasing distance translates into increasing time delays for radio transmission to and from ground control, thus requiring greater independence of the crew when handling health-related (and other) emergencies. Further, more autonomous health care will be required on long-duration, long-distance missions, where aborting the mission or evacuating a crew member for medical care is not a feasible option.

Another dimension of time involves the emergence of a predictable health problem related to the immediate mission task(s). Neurovestibular dysfunction, such as vertigo, might be an inconvenience during some parts of a mission, whereas neurovestibular dysfunction during the docking maneuver or a lunar or Mars landing could endanger mission success and crew welfare.

Finally, to be included meaningfully in the mission planning process, biomedical countermeasures and life support technologies must be validated well in advance of the final integrated mission plan, thus adding temporal urgency to the time dimension.

The committee notes that some of these dimensions of time are partially addressed by subdividing the risk analyses into the three Design Reference Missions in the BR, but concludes that time does not achieve the attention that is required to fully address risk priorities, determine countermeasure readiness, predict the maintainability of systems and equipment, and evaluate the impact of exploration missions flight on crew health. All of these factors deserve additional visibility in the BR.

Recommendation 2.1

The committee recommends that risk assessment and mitigation (technology or countermeasure development) in the BR be enhanced by labeling risks according to their relevance to operational requirements and to temporal urgency.

Interactions Among Risks

The BR currently contains various cross-cutting categories of risk, and an additional cross-cutting area is recommended later in this chapter. However, the BR does not provide enough information on the interrelations among and between these categories, risks, and their associated mitigation strategies. Space flight necessarily involves a complex system of interdependent systems, and as a result of that interdependence, any change that affects one system may also affect other systems, sometimes in ways that are neither obvious nor anticipated.

The BR contains a list of "related risks" and "interactions" (NASA, 2005, Appendixes A and B, respectively), but without explanation of how the risks were identified and assigned to specific categories. For example, nutrition is identified as a related risk for Risk 23 (human performance failure due to poor psychosocial adaptation) but not Risk 24 (human performance failure due to neurobehavioral changes). The extensive research literature supporting the role of nutritional factors in understanding the mechanisms of depression and other mental disorders that fall within Risk 24 (e.g., Young et al., 1985; Wurtman and Wurtman, 1989) is not cited. The BR provides no explanation that nutrition is related to poor psychosocial adaptation because the quality of food—as opposed to its nutritional content—is important in maintaining group morale under conditions of prolonged isolation and confinement (Stuster, 1996; Johnson et al., 2003).

Risks and their mitigations interact in a variety of complicated and sometimes subtle ways, and a comprehensive and continuous systems approach is required to anticipate their potential interactions and the impact of risk mitigation in one area on risk in an entirely unrelated area. Ideally, this systems approach should be applied at all levels, from the most minute components to the overall system in its broadest context, in order to anticipate and identify interactions among and between risks and risk mitigations.

An example of unanticipated interactions among risks is well illustrated by the interaction between risk mitigation for water contamination aboard the NASA orbiters and consequent thyroid dysfunction in crew members. Iodine was used as the bacteriostatic agent in drinking water aboard the U.S. orbiters—a seemingly reasonable approach to water purification. However, the concentration of iodine resulted in a daily iodine intake that far exceeded the recommended daily allowance and was sufficient to cause chemical evidence of thyroid dysfunction (e.g., increases in thyroid-stimulating hormone) commonly and clinical hyper- or hypothyroidism in several astronauts (IOM, 2004). Similar chemical or clinical abnormalities did not occur among those who flew on the Russian Mir space station even for intervals up to six months, but silver nitrate was used for water purification aboard Mir . (The problem was solved subsequently by installing an anion exchange resin filter at the water tap, but only after several years of thyroid abnormalities among flight crews.)

Other potential interactions appear in the BR. For example, many of the problems listed under Risk 25 (human performance failure due to neurobehavioral problems) are the result of problems listed under Risk 24 (human performance failure due to poor psychosocial adaptation (NRC, 1998; IOM, 2001). The poor coping skills of one crew member may result in reduced work productivity, leading to increased tension and conflict within the crew and resulting in reduced sleep and increased symptoms of depression, anxiety, and anger among all crew members (Palinkas, 1992, 2003). *Whereas some of these potential interactions are identified in Appendix B of the BR, that appendix is not referenced in the body of the text and the current version of the BR does not adequately emphasize these interrelations and their implications for countermeasure development, risk management, and risk mitigation.*

Recommendation 2.2

The committee recommends that greater effort be devoted to identifying and explaining the interrelations among risks and risk mitigations that are grouped within and across the cross-cutting categories in the BR.

Status of Readiness Levels

In general, the probability of an adverse event can be reduced in two ways: (1) by eliminating the risky procedure or activity or (2) through mitigation strategies and approaches that reduce the probability of adverse events to acceptable levels. Because space flight, in particular the exploration class missions, carries inherent risks, the BR emphasizes the latter approach through the development and application of countermeasures.

A variety of deliverable products result from the BR; they enable desirable outcomes or solutions to answer research and technology questions and reduce risk to support the human system in space. Progress in these areas is gauged by establishing "readiness levels" that delineate the level of maturity of countermeasures or technologies (Countermeasure Readiness Level [CRL] and Technology Readiness Level [TRL], Table 2-1). The process emulates safety improvement programs (U.S. Department of Trans-

TABLE 2-1 Countermeasure Readiness Level (CRL) and Technology Readiness Level (TRL)

TRL Definition	CRL DefinitionScore	TRL or CRL	CRL Category
Basic principles observed.	Phenomenon observed and reported. Problem defined.	1	Basic Research
Technology concept and/or application formulated.	Hypothesis formed, preliminary studies to define parameters. Demonstrate feasibility.	2	Research to Prove Feasibility
Analytical and experimental critical function/proof-of-concept.	Validated hypothesis. Understanding of scientific processes underlying problem.	3	
Component and/or breadboard validation in lab.	Formulation of countermeasures concept based on understanding of phenomenon.	4	Countermeasure Development
Component and/or breadboard in relevant environment.	Proof of concept testing and initial demonstration of feasibility and efficacy.	5	
System/subsystem model or prototype demonstration in relevant environment.	Laboratory/clinical testing of potential countermeasure in subjects to demonstrate efficacy of concept.	6	
Subsystem prototype in a space environment.	Evaluation with human subjects in controlled laboratory simulating operational space flight environment.	7	Countermeasure Demonstration
System completed and flight qualified through demonstration.	Validation with human subjects in actual operational space flight to demonstrate efficacy and operational feasibility.	8	Countermeasure Operations
System flight proven through mission operations.	Countermeasure fully flight-tested and ready for implementation.	9	

SOURCE: NASA (2005, Table 5-5).

portation, 2002), beginning with a problem definition and concluding with a defined operational improvement or countermeasure.

Currently, the BR proposes 183 projected deliverables. In the context of the BR, these deliverables constitute the risk mitigation plan for the human system. A more detailed description and estimation of readiness of each deliverable is provided in Appendix C of this report. The BR emphasizes, "Roadmap activities must focus on operational issues and solutions to operational problems to support an outcome-oriented approach." Thus, "bioastronautics research is focusing on deliverables at a readiness level of 4 or greater" (NASA, 2005, p. 17). *The committee concludes that this emphasis on an applied research agenda for NASA bioastronautics is not without significant consequences and risks, especially given the relatively immature status of current countermeasure development (see Figure 2-1).*

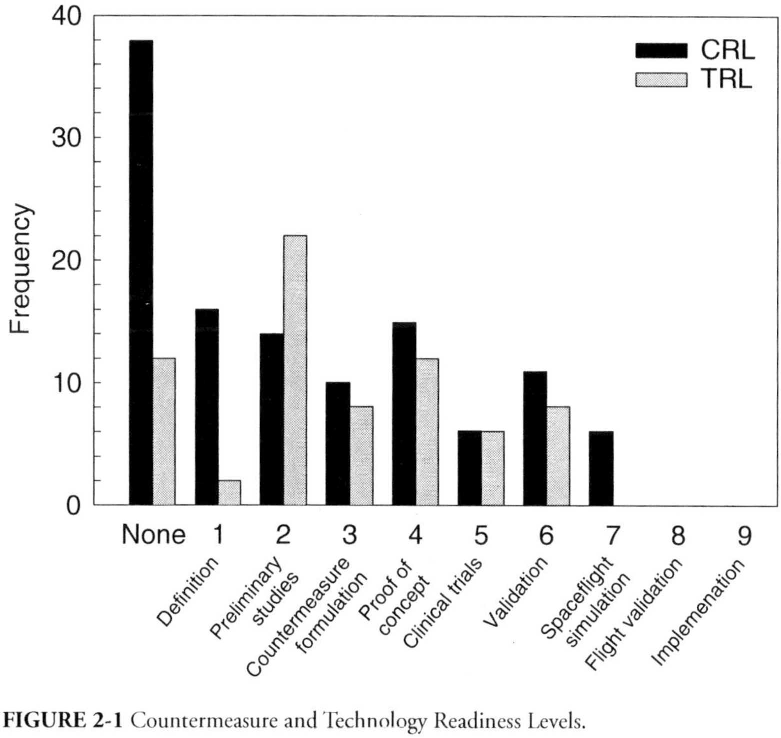

FIGURE 2-1 Countermeasure and Technology Readiness Levels.

Figure 2-1 summarizes the committee's analysis of the BR's proposed deliverables. An analysis of the forward work required to complete the BR risk mitigation plan reveals that 50 (27%) of the 183 proposed countermeasures and technology modifications are not yet defined. Of those that were defined, 71 (53%) are at the basic research stage of development (CRL or TRL levels 1–3), 56 (42%) are in the ground testing stage of development (CRL or TRL levels 4–6), and 6 (5%) have reached some stage of flight evaluation (CRL or TRL levels 7–9). *Thus, more than half of the defined deliverables proposed in the BR rank below the stage 4 level of readiness and, thus, below the threshold for priority in bioastronautics research consideration.* Included in these unranked or low-readiness areas are substantial portions of the mitigation plans on behavioral health and performance, radiation, autonomous health care, and water quality monitoring, areas that the committee finds deserving of further attention from NASA (IOM, 2005). To summarize, the state of countermeasure development significantly lags the need.

The committee finds that the majority of projected BR countermeasures, mitigations, or other deliverables are in a nascent state of readiness and are therefore unlikely to receive high-priority attention. Resources (described in Chapter 4) are unlikely to be sufficient to complete the BR mitigation plan in a time frame that enables the exploration class missions envisioned by NASA.

Establishing priorities for in-flight studies will be a significant challenge. Since NASA must address both near-term (e.g., the proposed lunar mission) and long-term (e.g., the Mars mission) objectives, the priorities for access to investigations aboard the ISS will require considerable wisdom to ensure that urgency is not confused with importance. Thus, some apparently "lower-priority" investigations may need to be manifested on the ISS earlier than others that might appear to be of higher priority. For example, studies regarding bone loss may have to take priority on the few remaining ISS flights in order to generate adequate in-flight data to support the Mars mission because countermeasures to bone loss will clearly be essential on a 30-month mission, even though that mission may be many years hence.

With the increasing likelihood that crew time, up-mass, and manning will continue to be problematic aboard the ISS, it will become increasingly important for surrogates of some sort to be used or developed to help assess the medical, nutritional, environmental, and behavioral issues that may confront astronauts who are in conditions of high stress in tight quarters for an extended period of time. The committee believes that analog environments and digital simulation will play an increasingly important role in the evaluation and mitigation of risks to the human in space and that the BR should reflect the importance of such analog approaches.

Clearly, analog environments can be created on Earth that could duplicate many of these conditions and thus could be used both as a test of hardware and to assess the quality of instruments used to create a selection process for crews. The value of such true analogs has already been demonstrated in bringing to light the medical problems associated with iodination of drinking water and the adverse effect that high iodine levels can have on crew members' thyroid function because this problem was demonstrated in ground-based simulations of space flight (IOM, 2004). Thus, specifically created mission analogs are useful to test not only behavioral issues but many of the medical issues that will face the crew as well. Practicality, however, would dictate that rather than "starting from scratch," the large amount of information that already exists from analog environments such as saturation diving, polar expeditions, bed rest, centrifuge, mock Crew Exploration Vehicle (CEV) expeditions, and submarines be used appropriately.

Submarines are but one example among the many analog environments that could be used both prospectively and retrospectively to help address some of the issues that may confront astronauts. From the 1960s to the mid-1980s, medical officers were assigned to submarines, and these physicians were required to write a research thesis on a relevant research topic. This resulted in a large number of unpublished but peer-reviewed theses, which can be obtained from the Naval Undersea Medical Institute (NUMI) and the Naval Submarine Medical Research Laboratory (NSMRL). A significant percentage of these reports addressed behavioral, environmental, and medical issues that were encountered on these deterrent patrols; therefore, they may be a useful resource for retrospective information.

Continuing the example in a prospective fashion, many parallels can be drawn between the "wardroom" of a nuclear submarine and a crew of astronauts on a long mission. Typically, wardroom officers on board a nuclear submarine are college graduates with highly technical educations, including postgraduate education. They are all highly motivated and committed to their jobs, and they live in confined environments for extended intervals. Thus, in several respects they are similar to flight crews, and they might be useful surrogates to examine the efficacy of various assessment instruments (e.g., for crew interaction, cohesiveness, and leadership). There are limitations, however, to the value of analog environments, based on the extent to which they reproduce the crew environment in the ISS or the proposed CEV. For example, the U.S. submarine service does not include women crew, nor does it usually include international crew as integral members of the wardroom team. U.S. experience might be supplemented by experience from other countries such as Norway and Denmark, whose submarines include women crew members, but this raises cultural differences as well.

Numerous other highly appropriate analog environments could be used. For example, the data from experimental saturation diving facilities could provide opportunities that mimic extravehicular activities, as well as crew quarters and environments; polar expeditions could provide information about long-term isolation; and prolonged bed rest can mimic at least several aspects of the musculoskeletal changes associated with microgravity. None by themselves is ideal, but collectively these and related approaches, supplemented by digital simulation approaches, can provide important supportive data in an era of restricted opportunities for in-flight clinical investigation.

Recommendation 2.3

The committee recommends that NASA initiate an aggressive program, including the use of animal models, analog environments, and space flight, to significantly accelerate the progress of all Countermeasure Readiness Levels and Technology Readiness Levels that are essential to support the proposed exploration agenda. Countermeasures and technologies at an undefined or low state of readiness (the majority of the current portfolio) should receive renewed attention. The committee notes further that failure to do so will jeopardize the exploration program outlined in the President's vision for exploration of January 2004.

FUSING THE RELATIONSHIPS BETWEEN HUMAN FACTORS AND TECHNOLOGY IN THE BR

The interrelationships and interactions between technology and human health and performance are evident in all aspects of daily life, but these take on new dimensions and importance as NASA moves to more ambitious exploration agendas, especially longer-term missions. Spacecrafts have evolved considerably to accommodate these needs; consider the evolution of internal volume and crew accommodations in the sequence of spacecraft from Gemini to Apollo to Orbiter to the ISS, for example, as longer-duration missions were planned. However, the consequences of human-technology interactions become more profound as mission duration and the expectations for crew task performance increase. (New space suits that permit crew to assemble fragile, possibly gossamer-like, components of radio telescopes or other devices while working in deeper space are an example; crew quarters for a 30-month mission are another, but innumerable examples could be cited in the current and future versions of the BR.) *The committee concludes that increased linkages in certain areas, specifically Human Systems Integration and Food and Nutrition, can be accomplished by modifying content in the BR.* Each one of these areas is addressed in the following sections.

Creating the Cross-Cutting Category "Human Systems Integration"

The BR lists six risks that correspond to human factors and behavior and performance. Four of these risks (23, 24, 25, and 26) fall within the cross-cutting category "Behavioral Health and Performance," and two of the risks (44 and 45) fall within the cross-cutting category "Advanced Life Support." These classifications reflect two separate perspectives, one based in the social, behavioral, and clinical sciences and one based on engineering and technology. However, both perspectives are critical to understanding human behavior and performance in space. These perspectives are also united by their treatment of the human both as a system of systems (e.g., bone, muscle, cardiovascular, respiratory, and neuropsychological) and as a component of a larger system of systems (e.g., advanced life support). Keeping these six risks in two separate cross-cutting areas may create certain limitations to understanding the linkages among these risks and, thus, may impair understanding of the interactions among risks and risk mitigation strategies.

As evidence of the linkage among risks that currently fall within two separate cross-cutting categories, both Risk 24 (human performance failure due to poor psychosocial adaptation) and Risk 45 (poorly integrated ground, crew, and automation functions) address ground–crew interactions. In Risk 24, the risks are framed in psychological terms such as displacement of hostility (Kanas et al., 2000), whereas in Risk 45 they are framed in technological terms (Caldwell, 2000, 2005). Although the two risks emphasize different causes of dysfunctional ground–crew relations, the causes are in fact interconnected to a greater extent than is reflected in the current separate classifications, and mitigation of these risks will likely require a combination of both engineering and psychological solutions. For example, poor engineering of information transmission routines within mission control, and between the crew and mission control, may impede ground–crew communication and decision making, resulting in increased ground–crew tension, increased hostility, and reduced cooperation in task completion (Kanas et al., 2000, 2001a,b). The committee noted that dysfunctional ground–crew communications have impacted previous missions, including quite publicly the ground–crew tension that was evident during Skylab 3 (Zimmerman, 2003, pp. 73–80).

The link between humans and engineering is essential for enhancing mission success and crew performance. For example, Callaghan et al. (1992) observed that the development of methods was as important as the development of standards. Similarly, Whitmore et al. (1999) indicated that the first research priority for the ISS should be in-flight research tools to understand habitability. The second priority was to identify human factors critical to habitability and determine how these could be tested in microgravity and analog environments. Adams (1999) also emphasized the need for habitability studies. Other researchers have highlighted the importance of human systems integration. Peacock et al. (2001), for example, identified the following technology issues for manned space missions: (1) limitations to astronaut mobility and force output while wearing extravehicular activity (EVA) clothing, (2) need for physical restraints and mobility aids in microgravity, and (3) maintaining spatial orientation. The need for enhanced human–robot interactions was identified by Gross et al. (2002), and Hartman (2003) identified the human systems interaction considerations associated with the martian weather, which includes temperature extremes (–87°C to –25°C) and windstorms that will cause dust to lodge in seams and crevices and reduce visibility to less than 1 meter at times, all in an atmosphere that is almost pure CO_2, with a barometric pressure that is 1 percent that of Earth's surface (about equivalent to the pressure at 110,000 feet above Earth).

The committee concluded that human factors are a high priority in space engineering design, especially in an era of planetary exploration. Linking human factors with engineering perspectives in the BR is essential for the development of countermeasures. For example, the extent of musculoskeletal weakness upon arrival in a gravitational environment after long-duration space flight will depend considerably on the engineering and technology (e.g., artificial gravity, exercise equipment) that are incorporated in the CEV. Similar arguments apply to the control of radiation exposure and the identification of coping skills and other preventive measures to address the human factors associated with prolonged space flight.

Recommendation 2.4

The committee recommends that NASA create a cross-cutting category that spans the two existing categories of "Behavioral Health and Performance" and "Space Human Factors Engineering" risks listed currently in the area of the "Advanced Human Support Technologies." This new category should be labeled "Human Systems Integration," consistent with the terminology currently in use by the U.S. Department of Defense (2004) and the Human Systems Integration section in NASA Headquarters. Further, the committee implies by this recommendation that NASA consider placing other risks in this category over time, to facilitate collaborative effort by clinical and engineering scientists in defining, measuring, and mitigating each of the Human Systems Integration risks.

Creating the Cross-Cutting Area "Food and Nutrition"

Maintaining the crew's overall health will be essential for the success of exploration missions, and food safety, food quality, and nutrition will play a major role in achieving overall health. Extensive reviews of the impact of nutrition on space travelers have been conducted (Stein, 2001; Lane and Feeback, 2002). The composition, safety, quality, and quantity of food and water may all impact the physical and mental health of the crew during space missions, particularly when such missions are long term.

Establishing the optimal nutrient requirements for the crew in space is critical not only for the maintenance of the crew's health during flight—the most immediate concern of the BR—but also for the health of astronauts after they return to Earth. However, there are very few data on the nutrient requirements of human subjects in extraterrestrial environments, and the problems posed by this scarcity of data are compounded by the small sample size from which existing data on the role of nutrition in maintaining human health in space are drawn. Thus, weight management and plasma levels of lipids, vitamins and minerals, salts, protein, and carbohydrate may be impacted by nutritional intake, in conjunction with the amount of physical activity undertaken by the crew and by space flight itself. Smith and Heer (2002) pointed to the need to ensure adequate dietary intake (and synthesis, when applicable) of various nutrients involved in bone and calcium homeostasis, including calcium, vitamins D and K, protein, sodium, and phosphorus. Similarly, homeostasis of red and white blood cells, as well as vascular integrity, bone mass, and osmolality, may be altered by the quality of nutrition (Smith et al., 2005), whereas crew morale can be affected by the appearance and quality of food (Kerwin and Seddon, 2002). In addition, adequate quantity and quality of water must be ensured.

Despite the efforts by NASA's nutrition and food scientists, there is evidence that astronauts have had energy intakes 30 percent to 40 percent lower than their needs during space missions (Smith and Lane, 1999), not because of faulty planning in terms of nutrient intake or unbalanced menus, but because of such factors as time constraints (e.g., time for meal preparation and consumption), work overload, and lack of appetite. Moreover, it is well established that astronauts' energy expenditures in space are unchanged or even increased from those on Earth (Lane et al., 1997; Stein et al., 1999).

According to Smith et al. (2005), several changes in nutritional status after space flight have been observed. In addition to insufficient energy intake, vitamin D levels were altered after long-duration missions (4 to 6 months), even when the vitamin was taken as a supplement during flight. Further, there was evidence that the metabolism of this vitamin may have been affected by the long-term stay of cosmonauts in the Mir station. Altered vitamin D status, in turn, was associated with increased bone resorption after landing. The same authors reported severe decreases of up to 45 percent for magnesium and phosphorus concentrations. The former may be of concern in long-term missions because of the role of magnesium in preventing formation of renal stones (Su et al., 1991).

Other vitamin levels are also affected by prolonged human exposure to the space environment. The level of vitamin K, an important element in maintaining bone health, was reported to decrease during space flight, and levels of vitamin E and folate may have decreased also (Smith et al., 2005). According to Smith et al., it is not known whether the observed decrease in folate—attributed to inadequate food intake, not to food processing techniques—would be accentuated during prolonged space missions. A review by Lucock (2000) suggests that the role of this B vitamin in maintaining good health may extend beyond prenatal conditioning and a positive influence on cardiovascular disease, to encompass several types of cancer, dementia, and affective disorders.

Decreased red blood cell (RBC) mass after short- and long-term space missions—at a rate of more than 1 percent per day, eventually reaching as high as 10 percent to 15 percent after 2 weeks of flight—is another concern reported and discussed by Smith (2002). He indicated that data from the Spacelab Life Sciences Shuttle missions showed that the release of new RBCs is decreased by weightlessness and that newly released RBCs, larger than the more mature circulating cells, are then selectively removed from the circulation. Such decreases in RBC mass were accompanied by decreases in body fluid volume, and both effects were attributed to the body's apparent sense of decreased blood volume requirement. As a result, anemia has been observed during many space missions. Destroyed RBCs, in turn, release iron, which is then stored, as evidenced by increased levels of sodium ferritin (Alfrey et al., 2000). Although indices of iron metabolism and erythropoiesis returned to normal within days of landing, increased levels of stored iron and its availability in tissue would suggest a need to reduce iron intake during long-duration missions, given that the long-term effects of this condition are unknown (Smith, 2002).

Muscle loss early in flight has been reported to occur as a result of reduced protein synthesis by inactive muscle (Ferrando et al., 1996) and to continue at a slower rate throughout space missions. As a result, astronauts returning from missions as short as 1 or 2 weeks have exhibited lack of coordination, weakness, and fatigue (Edgerton and Roy, 1994; Riley et al., 1995). These conditions, described by Adams et al. (2003) and reviewed by Baldwin et al. (2003), may adversely influence performance on extended missions.

In addition to nutritional intake, the space environment itself may have marked effects on the nutritional status of astronauts; the above-mentioned hematological changes during space flight are but one example. Another is the adverse effect of increased radiation in space, which may accelerate the degradation or oxidation of food components and alter the potential mitigation of these effects through the protective action of antioxidants (including various vitamins). Antioxidants could be added to food to protect both the foods and their intended consumers (Lane and Schoeller, 1999). As stated in a short, on-line version of these authors' work, "As long as space travelers can take food from earth with them, it may be possible to design diets that are capable of reducing these problems and other problems. However, as voyages become longer and longer, and the crews become more dependent on food that they will grow and process during their flight, this becomes more of a problem. It is already apparent that as the diet of astronauts becomes more vegetarian, plant sources rich in calcium and high in antioxidants with a high energy content will be most desirable" (Lane and Schoeller, 2005).

Finally, issues of palatability, variety, cultural preferences and expectations, and the social interactions and connotations of mealtime must be considered. Although food science and nutrition are two separate disciplines, their interrelation is so important that many universities across the nation and elsewhere—recognizing the need to more closely integrate these disciplines—have merged their food science and nutrition departments in recent years.

In view of the many facets of potential physiological and psychological impacts of nutrition on the crew during prolonged space missions, and the resulting effects on crew performance and physical and mental health during and after space missions, particularly long-term missions, the committee concurs that nutrition is a cross-cutting theme and should continue to be reflected as such in the BR. However, the committee emphasizes that nutrition alone does not encompass all of the concerns related to diet and believes that the title should be expanded to encompass the technology aspects of food safety and quality (Risk 38) and the clinical aspects of human nutrition (Risk 16) to further link the technology with the human factors, as recognized in universities around the world.

Recommendation 2.5

The committee recommends that the cross-cutting area "Nutrition" be renamed and expanded to "Food and Nutrition" to more fully encompass the concepts of dietary needs for space flight and emphasize the relationships between nutrition and food technology. Further, the committee recommends that the impact of inadequate food and nutrition on processes related to mental and physical health risks and maintenance of the space environment and life support systems be defined from the standpoints of food safety, quality, and quantity, and interventions proposed if problems occur.

SPECIFIC ISSUES

Specific issues are those areas that deserve focused attention and refinement in the BR in order to more effectively accomplish its goals as a management tool. They should be viewed as tactics for the BR, whereas the overall issues should be viewed as strategies.

Reclassification of Behavioral Health Risks

The existing cross-cutting category "Behavioral Health and Performance" in the BR includes four categories of risk: human performance failure due to poor psychosocial adaptation (Risk 24), human performance failure due to neurobehavioral problems (Risk 25), mismatch between crew cognitive capabilities and task demands (Risk 26), and human performance failure due to sleep loss and circadian rhythm problems (Risk 27). Unfortunately, there is considerable overlap in these categories. Both Risks 25 and 26 make reference to cognitive problems. Risks 24 and 25 both include mood disorders. Further, human performance failure due to poor psychosocial adaptation (Risk 24) belies the fact that there are three separate determinants or moderators of poor psychosocial adaptation under conditions of prolonged isolation and confinement (Palinkas, 2001). These include intrapersonal factors such as personality styles and strategies for coping with stress; interpersonal factors such as group cooperation and conflict; and organizational factors such as the cultural systems of the participating national space agencies and contractors, the values they embody, and their influence on health and behavior.

The countermeasures to address these problems are not well defined in the BR. Thus, reference is made to the use of screening and selection as a potential countermeasure to each of the four risks identified in behavioral health and performance. However, the BR does not acknowledge that there are at least two potential strategies for screening and selection. Screening and selection of astronaut personnel in the U.S. space program has traditionally been based on a "select-out" philosophy that excludes those with any diagnosable psychiatric disorder or high likelihood of developing such a disorder (Santy, 1994). Although this approach has generally been successful in minimizing decrements in behavior and performance during short-duration missions (1 to 14 days), the advent of longer-duration missions, ranging from earlier 3-month assignments aboard the Mir Space Station to the proposed 3-year mission to Mars, has generated greater interest in an approach that is based on a "select-in" philosophy. Such an approach would seek to identify candidates whose personality traits enhance their ability to function at high levels (both physically and mentally) while experiencing chronic stress and to interact productively as a member of a crew (Santy, 1994).

Another option would be to expand the range of mental disorders that are considered as part of the select-out criteria. People with personality disorders often make their own interpersonal problems worse because they are rigid, inflexible, and unable to adapt to the social challenges they face (Chen et al., 2004; Pagano et al., 2004), and they create problems for those around them as well. Some personality disorders, including narcissism (Morf and Rhodewalt, 2001), would be expected to disrupt group cohesion and interfere with interpersonal relations involving other members of the space crew as well as Earth-based support personnel.

The committee concludes that empirical evaluation of current (and proposed) select-in and select-out procedures must be conducted among members of the astronaut corps during all phases of training and mission preparation. At present, the BR offers no guidelines for improving select-out procedures to reduce the likelihood of human performance failure due to poor psychosocial adaptation, neurobehavioral changes, cognitive overload, or disruption of sleep and circadian rhythms. Additional screening criteria that might benefit from further research include history of family disorders, childhood experiences of abuse or trauma, and stress reactivity.

Two other approaches for screening should also be given further consideration in crew selection. One involves the method of selection of specific crew members from among the pool of eligible astronauts. At present, these decisions appear to be made with little consideration of an external professional evaluation of crew compatibility.

Recommendation 2.6

The committee recommends that the Astronaut Office and representative flight surgeons be consulted regarding the crew selection process in order to place greater emphasis on the roles of crew compatibility and team performance in overall mission success.

The second approach involves the use of analog environments and behavioral challenges to evaluate individual and team performance. After selection of a specific set of crew members and during their training for a mission, the crew should be assessed while it responds to carefully designed stressful experiences that would mimic events that could occur during a flight. These experiences might provide an opportunity to observe ways in which crew members respond, and these patterns might actually predict future performance more accurately than other kinds of assessment procedures. The committee recognizes that some of these concepts may be included in some of the research and technology questions in the BR, but they are not prominent and thus are pointed out here to emphasize that they should be part of the behavioral health planning process.

Similarly, the BR provides only modest guidance regarding incorporation of evidence-based interventions or countermeasures for mental disorders that might be experienced in space. There is evidence to support cognitive–behavioral treatments for depression or anxiety disorders in other populations (DeRubeis et al., 1999; Roy-Byrne et al., 2005), but their effectiveness in space flight or flight simulation is unknown. For example, brief forms of cognitive and interpersonal therapy have been demonstrated to be effective in the treatment of depression (Hollon et al., 2002; DeRubeis et al., 2005; Otto et al., 2005), but it is not clear that these interventions would be effective in a mode that did not allow for face-to-face discussions, or in which significant communication delays would impose extended pause times in the conversation. Careful evaluation of these procedures should be conducted using analog environments.

The committee concluded that the BR does not provide clear guidance for the research necessary to develop effective countermeasures for behavioral health risks of extended flight. In part, this is the result of a process that has "lumped" together certain distinct risks into the same category and, in part, due to "splitting" the same phenomena (e.g., mood disorders and cognitive performance) into more than one risk category without explication of their interrelationships. This has tended to both trivialize the risks and complicate efforts to develop effective countermeasures.

Recommendation 2.7

The committee recommends that behavioral health risks within the proposed new cross-cutting category of Human Systems Integration be grouped into categories based on clinical outcomes such as interpersonal conflict, affect regulation, decrements in cognitive performance, mood disorders, and sleep disorders, rather than categories such as psychosocial, neurobehavioral, cognitive, and circadian rhythms, and that the interrelations between these categories be delineated clearly.

The committee suggests convening a workshop of investigators, flight operations medical personnel, and expert clinicians who have experience in behavioral interventions for the management of mental disorders that develop during times of prolonged stress. The purpose would be to estimate the likelihood of such disorders developing during a 30-month extended mission, to identify the personality characteristics associated with a low risk for the development of such disorders, and to recommend evidence-based behavioral interventions to minimize the severity of the disorders if they do develop.

Finally, the committee notes that the BR contains no references to human sexuality, and this oversight should be corrected. Whereas the committee recognizes the task-oriented nature of both the crew and the mission, it concludes that ignoring the potential consequences of human sexuality is not appropriate when considering extended-duration missions. Areas of concern for the 30-month Mars mission include the potential psychological and physiological consequences of sexual activity, consequences that could endanger life, crew cohesion, performance, and mission success. Some risks can be eliminated but others cannot. For instance, the risk of pregnancy might be mitigated by crew selection. However, the long lead time to answer questions about the efficacy, safety, and side effects of contraceptive medications may require that studies to answer these questions be completed prior to crew selection or that other measures be used to mitigate the risk of pregnancy.

Recommendation 2.8

The committee recommends that issues of human sexuality be addressed in the BR in relation to long-duration missions such as the proposed Mars Design Reference Mission.

Psychological and Physical Impacts of Space Flight on Performance

To more fully address human behavior and performance issues during long-duration space flight, NASA must address a number of significant psychological and biological factors that will influence the ability of astronauts and crews to perform effectively. Effective performance includes the maintenance of individual high-level cognitive, communication, and physical skills and the maintenance of overall crew skills and teamwork. The latter require effective interactions with Earth-based support and recovery systems, as well as harmonious, flexible, and effective group relations among crew members over long periods of time and in an environment that may be associated with a variety of stresses.

Several of the risks associated with space flight involve human responses to stress. Stress-related performance decrements in extended space flight have been documented. Sauer et al. (1997) reviewed the evidence and postulated that sustained stress will result in a decrease in motivation and increased risk of errors. They stated that Soviet scientists reported decreased information processing speed within the first nine days of a space flight, especially for first-time crew members. Isolation studies conducted in the former Soviet Union using spacecraft simulators indicated that performance decrements included increased decision time, increased errors of commission, and decreased attention span. Cognitive effects associated with fatigue were also noted in hyperbaric chamber isolation studies. In actual extended space flights, fatigue increased over time, with associated increases in errors.

The physiological response to stress includes increased concentration of stress hormones (e.g., catecholamines, cortisol), and these alterations may affect a number of the risks identified in the BR (including Risks 5, 6, 8, 9, 10, 24, 25, 26, 27, and 28). Examples of altered human health associated with stress responses include cardiac dysrhythmias (Mackstaller and Alpert, 1997), enhanced bacterial growth (Klaus, 2002; Lyte, 2004), altered immune function with increased risk of infection (Rabin, 1999; Miller et al., 2004), exacerbation of allergies (Wright et al., 2005), and emergence or exacerbation of autoimmune disease (Ackerman et al., 2002). Further, stress hormone increases are associated with altered central nervous system function (Zarkovic et al., 2003; Brown et al., 2004), sleep disturbances, and altered circadian rhythms (Maschke and Hecht, 2004). Therefore, infectious disease, autoimmune disease, atherosclerotic heart disease, hypertension, and the mental conditions of depression and possibly accelerated loss of cognition may be influenced by the hormonal response to stress presented by an extended-duration space mission (McEwen et al., 1999). Although not all of these relationships have specifically been demonstrated during space flight, the prudent approach would be to assume that these relationships persist in space, at least until such time as this assumption is disproved (an unlikely possibility, given the limited opportunity mentioned earlier to collect significant data during flight).

Whereas many of these concerns are suggested in the BR, their specific links to stress hormones—and the implications thereof—are not emphasized, and neither are the implications for crew selection. The likelihood of maintaining mental and physical health during periods of prolonged space flight may be enhanced if crew members (1) have low responses to stress (low stressor-induced elevation of cortisol and catecholamines) and (2) are trained in coping skills that minimize the stressor-induced increase of these hormones. This would involve selecting astronauts who are low stress responders and who have good stress coping skills.

Another source of concern in long-duration space flights is the possibility that crew members might experience major mood swings related to either depression or mania at various times during the flight. For example, the etiology of clinical depression is undergoing reevaluation based on new studies indicating that increased levels of glucocorticoids can damage neurons that are important to the genesis of depression (Sapolsky, 2004).

Based on this and related evidence in the stress–response literature, the committee concluded that stress–response analyses should be included among selection criteria. Selection criteria that might be considered by NASA include low stress responsiveness determined by Stroop test measurements of blood pressure and heart rate (Probst et al., 1997; Seibt et al., 1998; McCaffery et al., 2003) and performance on psychological tests and personality questionnaires that have been demonstrated to be related to the ability to interact, communicate, and work effectively with others in a closed environment for extended intervals.

Radiation Effects: Establishing Risk-Specific Radiation Exposure Levels

Exposure to ionizing radiation from the solar wind, solar particle events (SPEs), galactic cosmic rays (GCRs) and other sources is a major concern for space flight beyond low Earth orbit (LEO). Complex deoxyribonucleic acid (DNA) damage from radiation may result in cell death, causing acute symptoms if enough cells in a tissue are affected. Also, DNA damage repair mechanisms (e.g., nonhomologous end joining) for complex damage are not error free, and repair errors can result in cellular mutations that contribute to eventual carcinogenesis. Whole-body radiation at doses high enough to cause serious acute effects has been associated with substantial—and probably unacceptable—increases in lifetime cancer risk in terrestrial populations (Hall, 2000). Therefore, radiation protection measures aimed at minimizing radiation-related cancer risk are also likely to protect against severe acute effects. A possible exception to this general observation is that central nervous system (CNS) tissue may be more vulnerable than other tissues, and CNS impairment may occur following exposure to high-mass, high-*Z*, high-energy (HZE) radiation from GCRs at low doses and low dose rates.

It is now well established that the radiation dose (mainly due to gamma rays) from the atomic bombings of Hiroshima and Nagasaki is also associated with increased risk for nonmalignant disease, particularly heart disease, stroke, respiratory disease, and digestive disease, and that the impact in terms of mortality may be roughly comparable to that seen for solid cancers (Preston et al., 2003). Radiation-induced cataracts are another established late health effect. Thus, any assessment of radiation-related health effects should include both malignant and nonmalignant disease.

The International Commission on Radiological Protection (ICRP) system of differential weighting of doses by radiation quality and by tissue—depending on susceptibility to radiation-related disease and the severity of such disease in terms of mortality and morbidity—may be a useful model for construction of a common metric for radiological risk (ICRP, 1991). (See also the successor document to ICRP 60, now in preparation; supporting documents and the report will be made available for public review on *www.icrp.org* prior to finalization.) However, the radiation environment of space travel beyond LEO is more complex, and less well understood, than the mostly low-dose radiation encountered by radiation workers and the general public that is the main concern of the broader radiation protection community. Also, individual tolerance for late health effect risks may be greater among the astronaut community than among the general population, although possibly not greater than that of some radiation workers. The NASA adaptation of the ICRP weighting scheme should be tailored to its unique exposure scenarios and should take more account of uncertainties in radiation dose and associated risks (i.e., it should address in a probabilistic way the question, How high could such risks plausibly be?).

Continuing experimental studies of biological effects from exposure to the radiation from GCRs—including experimental carcinogenesis—should refine our understanding of the relative biological effectiveness of such radiation compared to gamma and X-rays, so that the extensive dose–response information for these sparsely ionizing types of radiation can be adapted for application to risks associated with GCR exposure. As demonstrated by Cucinotta et al. (2001), it is possible to make meaningful inferences about cancer risk associated with exposure to GCRs based on expert judgment, with quantified subjective uncertainty estimates about the relative biological effectiveness of HZE radiation. Such inferences can and should be refined in the light of new research findings after these findings have been reviewed and evaluated by the wider radiation research community.

The amount of radiation exposure will have potential impact on the crew's health through increased risk of malignancy; altered hematopoieses due to bone marrow suppression; increased susceptibility to infection mediated by immune system suppression and possible mutation of microorganisms allowing them to escape from regulation by the innate host defense system; and unknown effects in tissue, such as CNS effects due to tissue damage and subsequent inflammatory response. Radiation therapy for patients with malignancy often has fatigue as a side effect (Jacobsen and Thors, 2003). Whether radiation exposure may pose a similar problem on long-term missions should be evaluated. Radiation risks differ considerably between a 12-month mission in LEO aboard the ISS and a 30-month mission to Mars because the latter will be exposed to significantly greater galactic cosmic radiation, for example.

Although the major technology issue is protecting astronauts from health risks associated with exposure to radiation, there are potential performance decrements associated with radiation exposure at lower levels than those that affect health. Astronauts traveling beyond LEO—particularly on lengthy missions—will have a high probability of exposure to GCRs, which are more numerous and energetic near the solar minimum, and to SPEs, which occur more frequently near the solar maximum. The effects of such exposures, especially those to GCRs, are not known, but some of these effects can be estimated with substantial (and quantifiable) uncertainty on the basis of present knowledge. As new experimental and theoretical information is accumulated, this uncertainty should decrease. According to Wilson et al. (2002), some protection against these effects may be afforded by either lunar or martian surfaces.

Adverse radiation effects on pharmacological preparations and food nutrients may also occur during missions. Some drugs are stable in a high-radiation environment, whereas others will not tolerate even moderate exposure to sunlight on Earth. Those currently deemed radiation-stable have not been studied in the unique HZE radiation environment characteristic of long-duration missions beyond LEO.

The committee concludes that radiation effects are given ample weight in the BR, and its suggestions are intended to draw NASA's attention to ways in which the BR might be refined, and perhaps simplified, when considering ra-

diation risks and risk mitigation. Further, the committee concurs with NASA's assessment that the duration of the mission and the distance from Earth will determine the amount and type of radiation exposure and the required shielding for both the vehicle itself and the protective wear for EVA, with the associated weight and design implications. However, the conventional rule of thumb for terrestrial radiation protection (i.e., that protection against late radiation effects such as increased cancer risk will also protect against acute radiation effects) may not hold for HZE from GCRs, particularly with regard to CNS impairment. Therefore, the committee also concludes that NASA must conduct further research to clarify the extent to which protracted or low-dose HZE radiation exposure might contribute to mission-damaging CNS effects.

> *Recommendation 2.9*
> **The committee recommends that a safe radiation exposure level be established by NASA for each relevant risk, based on projected flight duration and distance from Earth, and that the technology to keep the level of exposure below that limit be ensured.** Inherent in this recommendation, and consistent with Recommendation 2.3, the committee concludes that NASA must conduct further research to clarify the extent to which protracted or low-dose HZE radiation exposure might contribute to mission-damaging CNS effects.

As with the other components of the BR, it will be essential to follow the developments in both biological research and shielding technology to ensure crew health and safety for longer-duration, higher-radiation-exposure exploratory missions.

Assessing the Sources and Impact of Long-Duration Space Flight on Crew Health and Incremental Risk

Crew performance can be compromised by (1) intrinsic health alterations that occur spontaneously due to natural processes in the space environment, (2) aspects of the space environment that impair health, and (3) inadequate or malfunctioning life support systems. Differences among the Design Reference Missions (ISS, Moon, and Mars), in terms of factors such as duration, amount of gravity, and type and extent of radiation exposure, are examples of how the mission itself will influence the interactions between risks associated with natural processes, the space environment, and life support systems. *Thus, the risk priority and potential for interactions among these sources will vary based on the Design Reference Mission, and these factors will impact the severity, implications, and countermeasure development that would be appropriate for each situation. Whereas the BR relates the human health risks to the Design Reference Missions, it does not specifically categorize the etiology of the risk, and operational decisions may depend significantly on the source of these risks.*

Recommendation 2.10

The committee recommends that, wherever possible, NASA use actuarial data (such as those in the Longitudinal Study of Astronaut Health and the related comparison group of Johnson Space Center employees, as well as additional sources such a genomic data, where available) **to estimate and/or model the likelihood of intrinsic health alterations for crew who will be part of the Mars mission. Utilization of this information as part of the selection criteria for astronauts should be considered. After intrinsic health risks are estimated, NASA should then estimate and/or model the contribution of the space environment and life support system malfunction to increased risk.**

The committee notes that such approaches are used currently for the assessment of radiation risks and believes that the expansion of this concept will benefit NASA operations and decision making, as well as the astronauts, as they assess the risks of long-duration exploration missions. Because of the complexity of risk determination in the context of limited available research information, it may be useful also to look beyond traditional actuarial tables and utilize advanced computer simulations of human physiology to target mechanisms of risk and assist in the development of countermeasures. Such broad-based models of human physiology are readily available and have been used successfully to focus research design and delineate mechanisms. They can also be used in a predictive fashion when it is impossible to test the conditions through experimentation. A quantitative and integrative approach could be used to guide the BR process when objective data are unavailable or when sample sizes are unusually limited.

Autonomous Medical Care and Self-Care

The committee recognized that health care will, of necessity, be limited during space flight, and it will not be feasible to manage all medical conditions optimally during the three reference missions. The current status of CRLs clearly illustrates the magnitude of the problem for bioastronautics in general and for autonomous health care specifically. To prepare for risk mitigation and autonomous medical care to the extent possible, it may be helpful to consider the research issues in three categories, based on the type of problem and the nature of the research required to address the problem. These categories are the following:

- *Biological issues:* biological and pharmacological questions requiring basic research with cell cultures, animals, drugs, and so forth, in the space environment (most likely during the ISS and lunar missions).
- *Operational issues:* adaptations needed to make equipment and procedures operate effectively in microgravity or the environment of the ISS or exploration vehicle. Some of these studies can be conducted during parabolic flight with brief microgravity or simulated on Earth, whereas others will require validation in flight.
- *Health care delivery issues:* conditions that could be treated during a mission and the equipment and supplies required to implement that treatment. Some of these issues can be studied in analog environments; others may be explored by "thought experiments" with expert clinicians, on the basis of current and expected future development of novel diagnostic and therapeutic capabilities. Both approaches should be supplemented with digital simulations of acute responses and chronic adaptation that use models of contemporary research findings.

The BR addresses fundamental issues of wound and fracture healing and acknowledges that further studies must be done in the space environment. Anecdotal evidence suggests that lacerations and abrasions heal slowly in the space environment (Kirkpatrick et al., 1997). The complex interaction between the immune system, inflammation, and wound healing implies that alteration in any part of the system might affect the functioning of the system as a whole (Yang and Glaser, 2002). If a specific issue related to wound healing is identified, basic research to determine the nature of the deficit might yield an insight into the specific therapy. For example, high-dose corticosteroids impair wound healing, and this deficit is at least partially ameliorated by administration of vitamin A (Talas et al., 2002, 2003). Similar studies into fracture healing and the immune system in a prolonged space environment are outlined in the BR. Because of the likelihood that injuries will occur and the paucity of mitigation strategies at present, these and related studies must be given high priority. Similarly, the issue of drug stability in high-radiation environments mentioned earlier (see discussion of radiation) has to be addressed.

In the operational category, studies during parabolic flight have provided some guidance into issues such as airway management, cardiopulmonary resuscitation (CPR), control of body fluids (e.g., blood) in microgravity, and suction. A simple example of the operational challenges is apparent in the application of CPR: both victim and rescuer must be mechanically stabilized in order to deliver effective chest compression in microgravity, and there are few data that evaluate the effectiveness of CPR in providing organ perfusion under these conditions. Neither adequate suction—a basic requirement for airway management—nor the capability to vary inhaled oxygen concentration is currently available on the ISS (Bacal et al., 2004).

Both the biological and the operational research issues are aimed not at fundamental science, but at support of the specific health care delivery issues that are focused on crew health and mission success. What to treat? What not to treat? What to take in the vehicle's medical supply manifest? Precedents from the body of literature on health care rationing may be applicable to guide some of these health care delivery questions. An extensive review of the literature on health care in nonterrestrial environments, including a compilation of translated relevant Russian scientific literature, indicated that "the majority of resuscitative and surgical interventions required to stabilize a severely injured astronaut are feasible in a microgravity environment" (Kirkpatrick et al., 1997). However, the applicability of other health care techniques in such environments and the limitations imposed by upload volume and mass may preclude the availability of many techniques and impose a limited selection of options based on risk assessment and logistics. Cost–utility analysis should incorporate probabilistic modeling of the likelihood of encountering a specific adverse event and the benefit—both to the mission and to the individual—of taking action during the mission to mitigate the health problem (Sculpher et al., 2004; Seifan and Shemer, 2005).

As an example, what is the probability that a crew member will develop a malignancy during the mission? Breast cancer might serve as an example of these concerns. It is generally thought that breast cancer is diagnosed years after the first cells become cancerous and begin to multiply. Annual mammography, ultrasound, and physical examination still miss tumors that become apparent only in retrospect, and some women develop and present rapidly growing breast cancer between annual screening exams. The most sensitive diagnostic modality available currently is a magnetic resonance (MR) scan. However, even if crew were screened pre-flight with all of the available diagnostic modalities, it is conceivable that during the approximately 30-month duration of a projected Mars mission, a subclinical tumor might become clinically manifest. If the only treatment options were radical extirpative surgery, cytotoxic chemotherapy, focused radiation therapy, or some other complex and highly specialized modality, it appears unlikely that treatment would be implemented during the mission. This is an example of the time factor cited earlier. The longer the mission, the more likely is a clinically acute problem to surface. As previously mentioned (see the section on page 26 titled "The Time Factor and Its Impact on Risk"), mission duration will affect access to definitive medical care.

However, new diagnostic and therapeutic approaches may well alter current conclusions. The combination of more precise screening technologies and less invasive or debilitating therapies suggests that both diagnosis and treatment of early malignancies may be feasible in the future. Focused ultrasound is currently being used to seal bleeding vessels (Nields, 2005). Handheld diagnostic ultrasound units are a reality at present. Transcutaneous radio-frequency ablation is being used currently to eradicate small breast tumors (Wood et al., 2002). Thus it is conceivable that developments in ultrasound, MR, or some combination of focused energy diagnostic and therapeutic modalities may allow precise transcutaneous ablation of small tumors by the time a Mars mission is flown. Such a handheld unit would find other applications for other conditions such as treatment of trauma (Noble et al., 2002; Cornejo et al., 2004) and hence be a highly desirable addition to the medical supplies aboard the CEV. Under these circumstances, breast cancer—and perhaps some other malignancies—could move from the category of "cannot treat during mission" to the category of "pursue periodic screening and initiate treatment during mission." This example illustrates the benefits of focusing the mitigation strategies on the three areas described above and of using an iterative process to evaluate the current status of risks and mitigation strategies, as well as the importance again of linking technology, insight derived from models and simulation, and expert opinion when addressing risk mitigation in the BR.

A factor to be considered in deciding what to treat is that in the worst case, when Earth and Mars are most distant, communications can take up to 20 minutes one-way. Thus, any treatment regimen that necessitates real-time help from Earth would not be feasible, and the determination of what can and cannot be treated with a given crew and equipment mix must take into consideration that delays in communication may be as much as 40 minutes.

Crew selection will have a significant impact on health care delivery. For example, health care delivery approaches can be more sophisticated if a physician is selected as a crew member, and the reverse is true as well—a physician may be required if sophisticated care is deemed appropriate a priori. The committee noted that selection of an international crew could make the health delivery issues more complex because both expectations of and approaches to care differ among countries. Analog environment studies, quantitative risk–utility analyses, and multinational conferences may assist in resolving these questions, which must be addressed prospectively.

The committee concluded that it will be valuable to categorize health care risks into those with minimal and easily managed outcomes through those of increasing severity and decreasing capability for management due to complexity, distance, or duration of the mission in order to prioritize the biological, operational, and care delivery strategies related to the risks defined in the BR.

Recommendation 2.11

The committee recommends that a system be developed for quantitatively evaluating the mental and physical health risks that could affect mission success and crew health and that priorities for countermeasure development (i.e., definitive treatment vs. palliation) be established for the most likely conditions to be encountered during each reference mission. A panel of outstanding medical clinicians should be used to assist NASA medical operations staff in characterizing the likelihood, importance, and "treatability" of each condition.

REFERENCES

Ackerman KD, Heyman R, Rabin BS, Frank E, Anderson BP, Baum A. 2002. Stressful life events precede exacerbations of multiple sclerosis. *Psychosom. Med.* 64: 916–920.

Adams CM. 1999. The role of habitability studies in space facility and vehicle design. Houston, TX: Lockheed Martin Space Operations Company.

Adams GR, Caiozzo VJ, Baldwin KM. 2003. Skeletal muscle unweighting: spaceflight and ground based models. *J. Appl. Physiol.* 95: 2185–2201.

Alfrey CP, Rice L, Smith SM. 1999. Iron Metabolism and the changes in red blood cell metabolism. In Lane HW, Schoeller DA (eds.) *Nutrition in Spaceflight and Weightlessness Models.* Boca Raton, FL: CRC Press.

Bacal K, Beck G, McSwain NE Jr. 2004. A concept of operations for contingency medical care on the International Space Station. *Mil. Med.* 169(8): 631–641.

Baldwin KM, Edgerton VR, Roy RR. 2003. Muscle loss in space: physiological consequences. In Mark H, Salkin M, Yousef A (eds.) *Encyclopedia of Space Sciences and Technology.* Hoboken, NJ: Wiley & Sons, Inc.

Brown ES, Varghese FP, McEwen BS. 2004. Association of depression with medical illness: does cortisol play a role? *Biol. Psychiatry* 55: 1–9.

Caldwell BS. 2000. Information and communication technology needs for distributed communication and coordination during expedition-class spaceflight. *Aviat. Space Environ. Med.* 71(9 Suppl): A6–A10.

Caldwell BS. 2005. Analysis and modeling of information flow and distributed expertise in space-related operations. *Acta Astronaut.* 56: 996–1004.

Callaghan TF, Gosbee JW, Adam SC. 1992. Operational space human factors: methodology for a DSO (Paper 921156). Warrendale, PA: SAE International.

Chen H, Cohen P, Johnson JG, Kasen S, Sneed JR, Crawford TN. 2004. Adolescent personality disorders and conflict with romantic partners during the transition to adulthood. *J. Personal. Disord.* 18(6): 505–525.

Cornejo CJ, Vaezy S, Jurkovich GJ, Paun M, Sharar SR, Martin RW. 2004. High-intensity ultrasound treatment of blunt abdominal solid organ injury: an animal model. *J. Trauma* 57(1): 152–156.

Cucinotta FA, Schimmerling W, Wilson JW, Peterson LE, Badhwar G, Saganti P, Dicello JF. 2001. Space radiation cancer risks and uncertainties for Mars missions. *Rad. Res.* 156 (5): 682–688.

DeRubeis RJ, Gelfand LA, Tang TZ, Simons AD. 1999. Medications versus cognitive behavior therapy for severely depressed outpatients: mega-analysis of four randomized comparisons. *Am. J. Psychiatry* 156(7): 1007–1013.

DeRubeis RJ, Hollon SD, Amsterdam JD, Shelton RC, Young PR, Salomon RM, O'Reardon JP, Lovett ML, Gladis MM, Brown LL, Gallop R. 2005. Cognitive therapy vs. medications in the treatment of moderate to severe depression. *Arch. Gen. Psychiatry* 62(4):409–416.

Edgerton VR, Roy RR. 1994. Neuromuscular adaptation to actual and simulated weightlessness. *Adv. Space Biol. Med.* 4: 33–67.

Ferrando AA, Lane HW, Stuart CA, Davis-Street J, Wolfe RR. 1996. Prolonged bed rest decreases skeletal muscle and whole body protein synthesis. *Am. J. Physiol.* 270: E627–E633.

Gross AR, Briggs GA, Glass BJ, Pedersen L, Kortenkamp DM, Wettergreen DS, Nourbakhsh I, Clancy DJ. 2002. Advances in robotic, human, and autonomous systems for missions of space exploration. *Proc. 53rd International Astronautical Congress*, Houston, TX.

Hall EJ. 2000. *Radiobiology for the Radiologist,* 5th Ed. Philadelphia: Lippincott Williams & Wilkins.

Hartman WK. 2003. A Traveler's Guide to Mars. The Mysterious Landscapes of the Red Planet. New York: Workman Publishing.

Hollon SD, Thase ME, Markowitz JC. (2002). Treatment and prevention of depression. *Psychological Science in the Public Interest* 3: 39–77.

ICRP (International Commission on Radiological Protection). 1991. ICRP Publication: 1990. Recommendations of the International Commission on Radiological Protection, 60. Annex A. *Annals of the ICRP.* 21(1–3).

IOM (Institute of Medicine). 2001. *Safe Passage: Astronaut Care for Exploration Missions.* Washington, DC: National Academy Press.

IOM. 2004. *Review of NASA's Longitudinal Study of Astronaut Health.* Washington, DC: The National Academies Press.

IOM. 2005. *Preliminary Considerations Regarding NASA's Bioastronautics Critical Path Roadmap.* Washington, DC: The National Academies Press.

Jacobsen PB, Thors CL. 2003. Fatigue in the radiation therapy patient: current management and investigations. *Semin. Radiat. Oncol.* 13: 372–380.

Johnson JC, Boster JS, Palinkas LA. 2003. Social roles and the evolution of networks in extreme and isolated environments. *J. Math. Sociol.* 27: 89–121.

Kanas N, Salnitskiy V, Grund EM, Gushin V, Weiss DS, Kozerenko O, Sled A, Marmar CR. 2000. Interpersonal and cultural issues involving crews and ground personnel during Shuttle/Mir space missions. *Aviat. Space Environ. Med.* 71(9 Suppl.): A11–A16.

Kanas N, Salnitskiy V, Grund EM, Weiss DS, Gushin V, Kozerenko O, Sled A, Marmar CR. 2001a. Human interactions during Shuttle/Mir space missions. *Acta Astronaut.* 48(5): 777–784.

Kanas N, Salnitskiy V, Grund EM, Weiss DS, Gushin V, Kozerenko O, Sled A, Marmar CR. 2001b. Human interactions in space: results from Shuttle/Mir. *Acta Astronaut.* 49(3–10): 243–260.

Kerwin J, Seddon R. 2002. Eating in space—from an astronaut perspective. *Nutrition* 18: 921–925.

Kirkpatrick AW, Campbell MR, Novinkov OL, Goncharov IB, Kovachevich IV. 1997. Blunt trauma and operative care in microgravity: a review of microgravity physiology and surgical investigations with implications for critical care and operative treatment in space. *J. Am. Coll. Surg.* 184(5): 441–453.

Klaus DM. 2002. Space microbiology: microgravity and microorganisms. In Britton G (ed.) *The Encyclopedia of Environmental Microbiology.* New York: John Wiley & Sons, pp. 2996–3004.

Lane HW, Gratebeck RJ, Schoeller DA, Davis-Street J, Socki RA, Gibson EK. 1997. Comparison of ground-based and space flight energy expenditure and water turnover in middle-age healthy male US astronauts. *Am. J. Clin. Nutr.* 65: 4–12.

Lane HW, Feeback DL. 2002. History of nutrition in space flight: overview. *Nutrition* 18(10): 797–804.

Lane HW, Schoeller DA (eds.). 1999. *Nutrition in Spaceflight and Weightlessness Models.* Boca Raton, FL: CRC Press.

Lane HW, Schoeller DA. 2005. Cited: Nutrition challenges during space travel—solutions can benefit us all. *Medicinal Food News*. On-line [available: *http://www.medicinalfoodnews.com/vol06/issue3/space.htm*]. Accessed 5/05/05.

Lucock M. 2000. Folic acid: nutritional biochemistry, molecular biology, and role in disease processes. *Mol. Genet. Metab.* 71(1–2): 121–138.

Lyte M. 2004. Microbial endocrinology and infectious disease in the 21st century. *Trends Microbiol.* 12: 14–20.

Mackstaller L, Alpert J. 1997. Atrial fibrillation: a review of mechanism, etiology, and therapy. *Clin. Cardiol.* 20: 640–650.

Maschke C, Hecht K. 2004. Stress hormones and sleep disturbances—electrophysiological and hormonal aspects. *Noise Health* 6: 49–54.

McCaffery JM, Bleil M, Pogue-Geile MF, Ferrell RE, Manuck SB. 2003. Allelic variation in the serotonin transporter gene-linked polymorphic region (5-HTTLPR) and cardiovascular reactivity in young adult male and female twins of European–American descent. *Psychosom. Med.* 65: 721–728.

McEwen BS, de Leon MJ, Lupien SJ, Meaney MJ. 1999. Corticosteroids, the aging brain and cognition. *Trends Endocrinol. Metab.* 110: 92–96.

Miller GE, Cohen S, Pressman S, Barkin A, Rabin BS, Treanor JJ. 2004. Psychological stress and antibody response to influenza vaccination: when is the critical period for stress, and how does it get inside the body? *Psychosom. Med.* 66: 215–223.

Morf CC, Rhodewalt F. 2001. Unraveling the paradoxes of narcissism: a dynamic self-regulatory processing model. *Psychological Inquiry* 12: 177–196.

NASA (National Aeronautics and Space Administration). 2005. Bioastronautics Roadmap—a risk reduction strategy for human space exploration. On-line [available: *http://ston.jsc.nasa.gov/collections/TRS/-techrep/Sp-2005-6113.pdf*]. Accessed 1/6/2006.

Nields MW. 2005. Industry perspective: maximizing the benefit of improved detection with guided and monitored thermal ablation of small tumors. *Technol. Cancer Res. Treat.* 4(2): 123–130.

Noble ML, Vaezy S, Keshavarzi A, Paun M, Prokop AF, Chi EY, Cornejo C, Sharar SR, Jurkovich GJ, Martin RW, Crum LA. 2002. Spleen hemostasis using high-intensity ultrasound: survival and healing. *J. Trauma* 53(6): 1115–1120.

NRC (National Research Council). 1998. *A Strategy for Research in Space Biology and Medicine into the Next Century*. Washington, DC: National Academy Press.

Otto JW, Smits JAJ, Reese HE. 2005. Combined psychotherapy and pharmacotherapy for mood and anxiety disorders in adults: review and analysis. *Clinical Psychology: Science and Practice* 12: 72–86.

Pagano ME, Skodol AE, Stout RL, Shea MT, Yen S, Grilo CM, Sanislow CA, Bender DS, McGlashan TH, Zanarini MC, Gunderson JG. 2004. Stressful life events as predictors of functioning: findings from the collaborative longitudinal personality disorders study. *Acta Psychiatr. Scand.* 110(6): 421–429.

Palinkas LA. 1992. Going to extremes: the cultural context of stress, illness and coping in Antarctica. *Soc. Sci. Med.* 35(5): 651–664.

Palinkas LA. 2001. Psychosocial issues in long-term space flight: overview. *Gravitat. Space Biol. Bull.* 14(2): 25–34.

Palinkas LA. 2003. The psychology of isolated and confined environments: understanding human behavior in Antarctica. *Am. Psychologist* 58(5): 353 363.

Peacock B, Rajulu S, Novak J, Rathjen T, Whitmore M, Maida J, Woolford B. 2001. *Human factors and the international space station.* Proc. Human Factors and Ergonomics Soc. 45th Annual Meeting: 125–129.

Preston DL, Shimizu Y, Pierce DA, Suyama A, Mabuchi K. 2003. Studies of mortality of atomic bomb survivors: Report 13: Solid cancer and non-cancer diseases mortality: 1950–1997. *Radiat. Res.* 160: 381–407.

Probst M, Bulbulian R, Knapp C. 1997. Hemodynamic responses to the Stroop and cold pressor tests after submaximal cycling exercise in normotensive males. *Physiol. Behav.* 62: 1283–1290.

Rabin BS. 1999. *Stress, Immune Function, and Health: The Connection.* New York: John Wiley & Sons.

Riley DA, Thompson JL, Prippendorf B, Slocum GR. 1995. Review of spaceflight and hindlimb suspension unloading induced sarcomere damage and repair. *Basic Appl. Myology* 5: 139–145.

Roy-Byrne P, Berlliner L, Russo J, Zatzick D, Pitman R. 2005. Treatment preferences and determinants in victims of sexual and physical assault. *J. Nervous and Mental Disease* 191(3): 161–165.

Santy PA. 1994. *Choosing the Right Stuff: The Psychological Selection of Astronauts and Cosmonauts.* Westport, CT: Praeger Scientific.

Sapolsky RM. 2004. Is impaired neurogenesis relevant to the affective symptoms of depression? *Biol. Psychiatry* 56: 137–139.

Sauer J, Wastrell DG, Hockey GRJ. 1997. Skill maintenance in extended spaceflight: a human factors analysis of space and analogue work environments. *Acta Astronautica* 39(8): 579–587.

Sculpher MJ, Pang FS, Manca A, Drummond MF, Golder S, Urdahl H, Davies LM, Eastwood A. 2004. Generalisability in economic evaluation studies in healthcare: a review and case studies. *Health Technol. Assess.* 8(49): iii–iv, 1–192.

Seibt R, Boucsein W, Scheuch K. 1998. Effects of different stress settings on cardiovascular parameters and their relationship to daily life blood pressure in normotensives, borderline hypertensives and hypertensives. *Ergonomics* 41: 634–648.

Seifan A, Shemer J. 2005. Economic evaluation of medical technologies. *Isr. Med. Assoc. J.* 7(2): 67–70.

Smith SM. 2002. Red blood cell and iron metabolism during space flight. *Nutrition* 18: 864–866.

Smith SM, Heer MH. 2002. Calcium and bone metabolism during space flight. *Nutrition* 18: 849–852.

Smith SM, Lane HW. 1999. Gravity and space flight: effects on nutritional status. *Curr. Opin. Clin. Nutr. Metab. Care* 2: 335–338.

Smith SM, Zwart SR, Block G, Rice BL, Davis-Street JE. 2005. The nutritional status of astronauts is altered after long-term space flight aboard the International Space Station. *J. Nutrition* 135: 437–443.

Stein TP. 2001. Nutrition in the space station era. *Nutr. Res. Rev.* 14: 87–117.

Stein TP, Leskiw MJ, Schluter MD, Hoyt RW, Lane HW, Gratebeck RJ, LeBlanc AD. 1999. Energy expenditure and balance during spaceflight on the space shuttle. *Am. J. Physiol.* 276: R1739–R1748.

Stuster J. 1996. *Bold Endeavors: Lessons from Polar and Space Exploration*. Annapolis, MD: Naval Institute Press.

Su CJ, Shevock PN, Khan SR, Hackett RL. 1991. Effect of magnesium on calcium oxalate urolithiasis. *J. Urol.* 145: 1092–1095.

Talas DU, Nayci A, Atis S, Comelekoglu U, Polat A, Bagdatoglu C, Renda N. 2003. The effects of corticosteroids and vitamin A on the healing of tracheal anastomoses. *Int. J. Pediatr. Otorhinolaryngol.* 67(2): 109–116.

Talas DU, Nayci A, Atis S, Polat A, Comelekoglu U, Bagdatoglu C, Renda N. 2002. The effects of corticosteroids on the healing of tracheal anastomoses in a rat model. *Pharmacol. Res.* 45(4): 299–304.

U.S. DOD (U.S. Department of Defense). 2004. Defense Acquisition Guidebook. DoDD 5000.1 On-line [available: *http://akss.dau.mil/dag*].

U.S. DOT (U.S. Department of Transportation). 2002. National Review of the Highway Safety Improvement Program. Washington, DC: Federal Highway Administration. On-line [available: *http//www.tfhrc.gov/pubrds/02mar/04.htm*].

White House. 2004. President Bush announces new vision for space exploration program. Remarks by the President on U.S. space policy. On-line [available: *http://www.whitehouse.gov/news/releases/2004/01/20040114–3.html*]. Accessed 5/26/05.

Whitmore M, Adolf JA, Woolford BJ. 1999. Research priorities for the International Space Station. Houston, TX: NASA Johnson Space Center.

Wilson JW, Clowdsley MS, Cucinotta FA, Tripathi RK, Nealy JE, De Angelis G. 2002. Deep space design environments for human exploration. On-line. [available: *http://techreports.larc.nasa.gov/ltrs/PDF/2002/mtg/NASA-2002-wsc-jww2.pdf*]. Accessed 4/19/05.

Wood BJ, Ramkaransingh JR, Fojo T, Walther MM, Libutti SK. 2002. Percutaneous tumor ablation with radiofrequency. *Cancer* 94(2): 443–451.

Wright RJ, Cohen RT, Cohen S. 2005. The impact of stress on the development and expression of atopy. *Curr. Opin. Allergy Clin. Immunol.* 5: 23–29.

Wurtman RJ, Wurtman JJ. 1989. Carbohydrates and depression. *Sci. Am.* 260: 68–75.

Yang EV, Glaser R. 2002. Stress-induced immunomodulation and the implications for health. *Int. Immunopharmacol.* 2: 315–324.

Young SN, Smith SE, Pihl RO, Ervin FR. 1985. Tryptophan depletion causes a rapid lowering of mood in normal males. *Psychopharmacology* 87: 173–177.

Zarkovic M, Stefanova E, Ciric J, Penezic Z, Kostic V, Sumarac-Dumanovic M, Macut D, Ivovic MS, Gligorovic PV. 2003. Prolonged psychological stress suppresses cortisol secretion. *Clin. Endocrinol.* 59: 811–816.

Zimmerman R. 2003. *Leaving Earth: Space Stations, Rival Superpowers, and the Quest for Interplanetary Travel.* Washington, DC: Joseph Henry Press.

3

Considerations Regarding the BR Process

Several issues contribute to the complexity of the risks that the Bioastronautics Roadmap (BR) addresses, including the number of identified risks, the heterogeneity of risk types, and the interdependence among risks. In addition to the challenge of complexity, the determination of risk regarding activities for which there is little or no operational experience always involves an element of uncertainty, and the degree of uncertainty should be incorporated into the BR if it is to be a useful tool for decision making. Once identified and characterized, the method by which risk-related information is stored and made available to users also affects the overall effectiveness of the BR. This chapter reviews and comments on the methods used to construct the BR and on the format and approach used to represent and communicate its content.

RISK ASSESSMENT

As noted in the introduction to this report, efforts to understand and manage the risks associated with human space flight have been ongoing at the National Aeronautics and Space Administration (NASA) for many years, and specific activities related to the development of a roadmap began in the early 1990s. The process of risk identification that resulted in the BR commenced in 1997, in brainstorming sessions involving NASA, the National Space Biomedical Research Institute (NSBRI) staff and collaborators, and non-NASA experts who rated risks within their own discipline areas. With guidance from National Academies' (see Appendix A) and other advisory reports, 150 risks were identified. More recently, and after several iterations, the list was culled to the 45 risks that are the focus of the current BR (*http://bioastroroadmap.nasa.gov*).

The current set of risks and related research and technology questions were identified through an iterative process that included input from the discipline teams, the Bioastronautics Science Management Team, the Chief Health and Medical Officer, the Astronaut Office, flight surgeons, and NASA research management.

In the characterization of risks contained in the BR, risk assessment criteria included the determination of the likelihood of occurrence; the severity of consequences of each risk in terms of the crew's health, safety, and ability to perform mission objectives; and the state of the mitigation strategy. Relative risk priorities were derived from that assessment. Each risk has a set of associated research and technology questions. The answers to these questions are intended to lead to (1) risk assessment and quantification, (2) the development of countermeasures to prevent or mitigate the deleterious effects of space flight, (3) an improved basic understanding of underlying processes, and (4) medical diagnostic and treatment capabilities. This risk-based approach was devised to enable the development of a more rigorous decision-making process for allocation and implementation of resources, risk prioritization, access to facilities, operational requirement implementation, and crew time, as well as for development of cost-effective countermeasures and the design and implementation of effective advanced life support technology.

The committee agrees that NASA's decision to draw on expert opinion in identifying and ranking risks is a reasonable strategy, given the broad array of topics addressed in the BR and the need, in some cases, to characterize risks by extrapolation from current experience. However, the committee believes there are weaknesses in the current risk assessment process related to (1) lack of information regarding the quality of evidence that informs risk assessment, (2) the obscuring of risk that results from "lumping" both the risk and its associated mitigation into a single value, and (3) lack of a quantitative measurement of uncertainty related to the risk. These areas can, and should, be enhanced, and they are discussed in the sections leading to Recommendations 3.1, 3.2, and 3.3. These recommendations flow from an understanding that expert opinion in health care and the life sciences is influenced both by systematically derived data and by heuristics, or "rules of thumb," that are derived from personal and group experience

(McDonald, 1996) and an awareness that the risks in the BR range from theoretical concerns (e.g., virus-induced lymphomas and leukemia) to practical issues (e.g., nutrition, motion sickness, and bone and muscle loss). In addition, some risks are specific (e.g., renal stone formation), whereas others are general (e.g., ambulatory care).

The committee believes that risk assessment should primarily be evidence-based wherever possible, and where evidence does not exist, research should be directed to acquiring the evidence needed (see example on bone fracture risk associated with prolonged exposure to microgravity, Appendix D). Rating the quality of each published source of information should be a component of the BR. Quality ratings should follow commonly used criteria such as those of the Cochrane Collaboration (Starr and Chalmers, 2005) or the Agency for Healthcare Research and Quality (AHRQ, 2002) National Guideline Clearinghouse, recognizing that quality rating methods based only on the published literature will have to be modified to accommodate other sources such as conclusions from workshops and planning meetings. For risks that have more than one information source, the quality of the most robust source should be noted.

Recommendation 3.1

The committee recommends that NASA determine, and incorporate into the BR, measures of the quality of the evidence that forms the basis for defining risks and the assessments associated with each risk.

Disaggregation of Risk from Mitigation of Risk

Currently, the status of risks is represented in the BR as the aggregation of the importance of the risk (probability of occurrence × severity of consequences) combined with the effectiveness of the current state of countermeasures. In this model, a "severe" risk to the crew—such as loss of breathable atmosphere—would be represented as being minimal if the state of life support technology readiness or life support capability readiness to mitigate the risk was high. The committee notes that it is important to explicitly and *separately* assess and represent the importance of each risk and the state of the mitigation strategy or countermeasures that address the risk.

Disaggregation of mitigations is important for several reasons. First, it is important because aggregated risk values obscure the risk itself and can lead to false confidence or concern, depending on the status of countermeasure or technology development. Second, the functioning of many parts of the space flight system is closely linked to the functioning of other parts of the system, and approaches that are used to mitigate one risk may have a positive or negative impact on other risks or mitigations. Finally, as the space flight system evolves, changes in systems may have a positive or negative impact on the mitigations. Failure to track risk separately from risk mitigation could well lead to failure to focus on the inherent relationships among risks and mitigations. For example, the risk of high "*g*'s" for a fighter pilot can be mitigated adequately by a "*g*-suit" system that is inflated by air pressure, and disaggregation of the mitigation from the risk helps ensure that system designers are aware that any changes that may affect the ability of the system to inflate the *g*-suit could affect the mitigation of the risk of *g*-induced incapacity. The committee observes that this disaggregation will also help maintain a clear understanding that the notion of "retiring risks" due to the availability of effective countermeasures is seldom an accurate depiction of the state of operational readiness, since the underlying health and safety issues (e.g., loss of breathable atmosphere) remain a concern and are not "retired" by the existence of a life support system.

Recommendation 3.2
The committee recommends that NASA structure the BR to represent separately the severity and likelihood of each risk and the state of the mitigation strategy or countermeasures associated with each risk.

Expressing Uncertainty

The determination of risk always involves an element of uncertainty. To fully communicate the likelihood of occurrence of an event, it is necessary to communicate the extent of uncertainty in the assessment. The uncertainty associated with a risk may be represented by objective measures such as statistical confidence intervals and sensitivity analyses (NCRP, 1996, 1997; Warren-Hicks and Moore, 1998; Grogan et al., 2000; NCI–CDC, 2003) or by less objective but potentially useful techniques such as approximate reasoning (Hayes et al., 1979) or possibility theory using fuzzy sets (Zadeh, 1978). It is also possible to state uncertainty using narrative descriptions of the risk, such as expert opinion obtained in focus group settings (Cacuci, 1988; Lash and Silliman, 2000). The current printed and on-line versions of the BR do not include any expression of uncertainty in terms of risk estimates, reported confidence intervals, or narrative discussion.

Variability of Opinion About BR Risks

Heuristic solutions result from an informed set of principles or rules that an individual or group uses for decision making. Although such rules may rely on an empirically derived knowledge base, it is important to understand that experience, judgment, personal philosophies, and external pressure can affect a heuristic, causing different groups to reach different conclusions. Given that it is formulated from a set of risks derived from discussions among different teams of disciplinary experts, the BR is not immune to such effects. However, ongoing discussions among the Bioastronautics Science Management Team, the Office of the Chief Health and Medical Officer, the Astronaut Office, flight surgeons, and research management have led to views of the risks of space flight that are generally similar, but not identical, within and outside NASA.

For example, the Institute of Medicine (IOM, 2001) panel that authored *Safe Passage: Astronaut Care for Exploration Missions* selected four risks that earned the panel's highest rating of "severe" for a flight to Mars:

1. Trauma and acute medical problems
2. Carcinogenesis caused by radiation
3. Human performance failure because of poor psychosocial adaptation
4. Acceleration of age-related osteoporosis

The version of the BR originally supplied to the IOM for review (NASA, 2004) used similar language, listing the most serious risks for a Mars mission as the following:

1. Addressing the requirements for autonomous medical care
2. Providing radiation protection
3. Maintaining behavioral health and performance
4. Bone loss–related issues
5. Advanced human support technology

Flight surgeons and astronauts provided the BR review panel with a narrower, operationally focused set of priorities for exploration (Baker et al., 2004). Analogous sentiments are echoed in the most recent version of the Bioastronautics Roadmap (NASA, 2005), which states that "actual risks must be operationally based, not research based." Flight personnel identified four high-priority items:

1. Medical care diagnosis and treatment
2. Radiation exposure
3. Behavioral health and performance
4. Neurovestibular functionality

Bone loss was considered a lesser, but still important, item. The difference between this highest-priority list and preceding ones is that flight personnel place higher value on the ability of an astronaut to pilot a spacecraft during nominal and/or contingency operations. At the same time, they expressed the belief that successful countermeasures for musculoskeletal deconditioning are imminent.

An excellent example of the different values applied to the BR comes from the results of a consensus workshop held in May 2004 in Houston. Representatives of the Astronaut Office, Space Medicine Office, and the Bioastronautics Science Management Team (BSMT) met to review the BR and assess its suitability for reducing the risk of human space travel to Mars. Participants were asked to rate their concern about the subset of BR human risks, the need for future research in each area, and modifications to any of the risk set. Figure 3.1 summarizes data supplied to the committee.

Although the median "worriness" rating for the high risk category in the BR exceeded that of the medium risk category (median values 6.0 vs. 2.5; $p = 0.015$), the considerable overlap of "worriness" scores among the high, medium, and low risk BR categories (dark, medium and light gray dots, respectively in Figure 3-1) suggests that different heuristics to evaluate risk for human space exploration operate within NASA. Those communities most directly affected by the risk of a human space flight to Mars (i.e., astronauts and flight physicians) did not share all concerns raised by the disciplinary experts responsible for developing the BR, and in several cases felt that some risks did not belong in the Bioastronautics Roadmap (i.e., "worriness" score of 0).

Recommendation 3.3

The committee recommends that whenever possible, NASA restructure the BR to include a quantification of the uncertainty

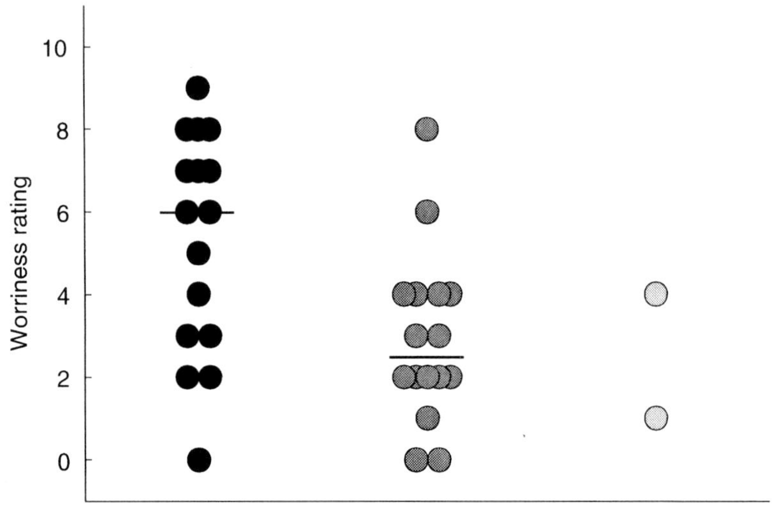

FIGURE 3-1 Results of the CB/SD/BSMT consensus workshop held May 25–26, 2004, in Houston, Texas. Representatives of the Astronaut Office, Space Medicine Office, and NASA bioastronautics management were asked to formulate a consensus rating of each Bioastronautics Critical Path Roadmap (BCPR, currently Bioastronautics Roadmap) risk on a "worriness" scale in response to the question, How worried would you be about this risk if we were to go to Mars today? A score of 0 indicated no concern; a score of 10 represented the greatest possible concern. Shaded dots (black, medium gray, light gray) represent the rating of human health risk (high, medium, or low, respectively) in the BR at the time of this assessment (May 2004). Horizontal bars indicate median rating.

of risks using standard uncertainty analysis techniques (e.g., frequentist, Bayesian, or possibility theory and approximate reasoning) that will provide uncertainty distributions or ranges in addition to point estimates. This will help contribute to the subsequent definition of operating bands.

The operating bands and exposure limits coming from the BR can be used as the basis for integrating bioastronautics risks into full-scope risk analyses for each of the reference missions. These larger mission risk analyses should be scenario-based, quantitative risk analyses. This approach would allow the evaluation and prioritization of all types of risk within a common framework.

DETERMINATION OF ACCEPTABLE RISK

The committee believes that the BR can be a tool to support both identification and *management* of risks such that each risk reaches an "acceptable" status for the relevant mission by the intended launch time for that mission. Several potentially useful risk management systems currently exist (e.g., NASA Continuous Risk Management Program [NASA, 1999], U.S. Navy Virginia Class Submarine [Kulez, 2003]). The committee does not recommend a specific implementation system but observes that a roadmap that supports risk management will have to contain elements that support operations in addition to those that point to needs for further research.

In this regard, although research in most fields may continue ad infinitum, the BR should attempt to identify "what is good enough" for the launch of a given category of mission. Researchers in virtually all fields are reluctant to declare total success, since this would be tantamount to forfeiting future funding. In the conduct of exploration, leaders cannot wait until every detail is resolved definitively, but only until the collective risk is mitigated adequately or otherwise reduced to permit a high enough level of optimism to justify mission initiation. This by no means suggests that research in a field should be terminated when sufficient progress has been made for launch, only that the mission should be "cleared" and further research dissociated from the operational aspects of the mission.

The presentation of information relative to risk management will have to be linked to the Design Reference Missions. The three different missions under consideration—1 year aboard the International Space Station (ISS), 1 month on the lunar surface, and a 30-month mission to Mars and back—will have different detailed risk management requirements in each risk category. A risk currently considered unacceptable for the Mars mission may very well be acceptable for the ISS mission. The fact that four Russian individuals (Titov, Manarov, Polyakov, and Avdeyev) have already spent a continuous year or more each in low Earth orbit suggests that the ISS mission category should be rephrased to either (1) identify additional detailed (scientific) mission objectives over and above mere survival for a year in orbit or (2) call for qualifying humans for routine and/or repetitive mission durations of 1 year in the ISS.

RISK COMMUNICATION

The manner in which risk-related information in the BR is represented and communicated to users is important to its overall effectiveness as a program management tool. The committee has identified issues of representation and communication that would help support the full range of BR stakeholders, notably NASA medical operations personnel and astronauts, in addition to NASA-funded researchers and contractors engaged in countermeasure development. *Central among these issues is that the BR has been presented as a document for review. The committee believes that the BR should be thought of and designed as a dynamic database of information relative to risk definition and assessment from which a document or set of alternative documents can be derived at any point in time and incorporated into a risk management program.* Databases lend themselves to the creation of multiple, different views of subsets of information in the database, and this capability parallels the need for many different groups of stakeholders to view the overall BR from different perspectives and at different levels of detail. Viewing and structuring the BR as a database will also facilitate keeping it current as new knowledge and new technologies emerge. The web-based on-line version of the BR—with its searchable interface and alternative views for aggregating BR information—is an important step in this direction.

The current BR contains for each risk a section titled "Important References," and many of these include hypertext links to the citation with abstract. It is not possible, however, to determine which references are linked to the specific elements of the BR structure (e.g., risk description, risk rating, current countermeasures, research and technology questions). The availability of links to citations is helpful and would be enhanced by links to the full text of publications, wherever feasible and within the constraints of copyright law.

A variety of alternative methods for scoring and communicating risk information may help with communication to different BR user communities. One widely used format for representing risks not currently incorporated into the BR is the NASA-wide Continuous Risk Management Program (NASA, 1999), and the committee encourages its continued use because it is widely recognized and understood throughout NASA. NASA developed the Continuous Risk Management Program in 1996 to help project managers continuously identify, analyze, and manage risk throughout the life cycle of a project and for use as a proactive tool for managers to monitor resource allocation and ensure that critical project milestones are achieved within an acceptable level of risk. The use of the Continuous Risk Management Program results in a set of actionable risks that can be assessed with regard to the probability and consequences of occurrence. This information could be used to plan mitigation measures indicating that all risks have been reduced to acceptable levels by the projected launch date, to inform cost–benefit analyses and prioritization efforts, and to help NASA obtain adequate resources (funding, time, expertise) to carry out these measures. Representation of BR risks in this format, in addition to the current formats, may be an effective supplemental way of communicating the elements of the BR throughout the organization.

Recommendation 3.4

To enhance effective communication of the content of the BR, the committee recommends that the BR be designed and utilized as a dynamic database of information relative to risk definition and assessment, from which a document or set of alternative documents can be derived at any time and incorporated into a risk management program.

KEEPING THE BR CURRENT

The period of time over which the Design Reference Missions will be planned and executed is decades. Thus, it is fundamentally important that configuration control methods be established and implemented for keeping the BR up-to-date as new knowledge and technologies develop. In this regard, the committee observed that new literature relative to risks in the BR that became available over the course of committee deliberations has not yet been incorporated into the BR. A mechanism is needed for periodic searches of the literature for information related to risks—including research conducted in space or in analog environments—as well as literature on the status of validation of existing countermeasures. This process can be facilitated by identifying an "owner and manager" within NASA for each set of related BR risks and establishing a regular review cycle. Given the rate of publication of new literature, it seems prudent to conduct reviews for updating not less than once annually.

Where there is a desire to combine published research data with "expert opinion" from stakeholders, methods such as computer modeling (White et al., 2003), Bayesian updating, and elicitation of expert opinion are available (see Appendix E for a description of the Bayesian update process). When relevant, digital modeling and simulation should be used to integrate and extrapolate findings. Regardless of the combination of research data and expert opinion, the long time frame of the space initiative makes it imperative that new knowledge and technologies be incorporated into the BR.

Recommendation 3.5

The committee recommends that the references cited in the BR be updated to reflect more recent research on both risk identification and countermeasure development. Moreover, a mechanism should be established for the *ongoing review of the current best evidence* contained in the research literature, and methods should be developed to integrate new findings from the literature with the expert opinion of key stakeholders, including those from operations and the research community.

INDEPENDENT HEALTH AND MEDICAL AUTHORITY

The Columbia Accident Investigation Board's (CAIB's) report recommended the establishment of an Independent Technical Engineering Authority "that is responsible for technical requirements and all waivers to them" in order to "build a disciplined, systematic approach to identifying, analyzing, and controlling hazards through the life cycle of the Shuttle System" (CAIB, 2003). The Office of the Chief Engineer has now been established as the NASA Independent Technical Authority (ITA), whose mission includes the following: developing and maintaining technical standards; serving as the sole waiver-granting authority for technical standards; conducting risk analyses; serving as owner of the failure mode, effects analysis, and hazard reporting systems; deciding what is and is not an anomalous event; and independently verifying launch readiness.

The committee finds that the decisions that will have to be made with respect to health and human safety aspects of the Design Reference Missions are similar in complexity to those relating to the Shuttle as a system of systems. Additionally, there may be a number of human health issues for which there are not sufficient data to perform a definitive risk analysis or perform conclusive testing or simulation prior to the mission, and there may be a need to accept a substantial amount of unmitigated risk to the crew. It seems reasonable to conclude that establishment of an Independent Health and Medical Authority (IHMA) with a scope of authority and responsibility analogous to the ITA, but focused on matters of crew health, safety, and effectiveness, is an appropriate management construct within the agency. Just as the ITA is the owner of technical standards, risk analysis, waiver granting, and verifying launch readiness for systems, an IHMA would do the same for human health and medical care issues. The involvement of the IHMA in assessing the ongoing activities of the BR—especially as they change and evolve over time—would foster a "closed-loop" approach for monitoring progress in risk reduction in each of the BR risk areas. The IHMA could delegate authority appropriately through a system of warrant holders such as that used by the ITA, with cooperation between discipline experts in the development and maintenance of standards, especially in areas of potential overlap such as environmental monitoring and human factors. Warrant holders would be expected to coordinate closely with each other when deviances to standards are considered.

Recommendation 3.6

The committee endorses the principle that critical decisions regarding safety and health should be made by an authority that is independent of programmatic costs and schedules. Given the importance and complexity of health and human safety issues, the committee acknowledges and endorses the creation of the Independent Health and Medical Authority (IMHA), analogous to the Independent Technical Authority, and recommends that the IHMA be given responsibility, authority, and accountability for the health and safety decisions that relate to risks identified in the BR.

REFERENCES

AHRQ (Agency for Healthcare Research and Quality). 2002. *Systems to Rate the Strength of Scientific Evidence.* Summary, Evidence Report/Technology Assessment: Number 47. AHRQ Publication No. 02-E015. Rockville, MD: Agency for Healthcare Research and Quality. On-line [available: *http://www.ahrq.gov/clinic/epcsums/strengthsum.htm*]. Accessed 7/25/05.

Baker E, Barratt M, Duncan M, Clark J, Cintron N. 2004. Operational Priorities for Medical Research for Exploration. Presentation to committee, NASA/Johnson Space Center, Houston, TX, November 17.

Cacuci DG. 1988. The forward and the adjoint methods of sensitivity analysis. Ch. 3 in Ronen Y (ed.) *Uncertainty Analysis,* Boca Raton, FL: CRC Press.

CAIB (Columbia Accident Investigation Board). 2003. Columbia Accident Investigation Board Report . On-line [available: *http://www.caib.us*]. Accessed 4/18/05.

Grogan HA, Sinclair WK, Voilleque PG. 2000. Assessing Risks from Exposure to Plutonium. Final Report. Part of Task 3: Independent Analysis of Exposure, Dose and Health Risk to Offsite Individuals. Radiological Assessment Corporation (*RAC*) Report No. 5, Revision 2, February. Neeses, SC.

Hayes J, Michie D, Mikulich LI (eds.). 1979. *A Theory of Approximate Reasoning, Machine Intelligence.* New York: Halstead Press.

IOM (Institute of Medicine). 2001. *Safe Passage: Astronaut Care for Exploration Missions.* Board on Health Science Policy, Institute of Medicine. Washington, DC: National Academy Press.

Kulez J. (September 4-5, 2003). *Risk Integration Approaches and Results for the Virginia Class Submarine.* [Online]. Available: http://www.atc.nasa.gov/hostedEvents/rmc4/presentations/Day%201%209-4-03/3%20pm%20Kulesz.ppt#5 [Accessed January 10, 2006].

Lash TL, Silliman RA. 2000. A sensitivity analysis to separate bias due to confounding from bias due to predicting misclassification by a variable that does both. *Epidemiology* 11(5): 544-549.

McDonald CJ. 1996. Medical heuristics: the silent adjudicators of clinical practice. *Ann. Intern. Med.* 124(1 Pt 1): 56-62.

NASA (National Aeronautics and Space Administration). 1999. NASA Continuous Risk Management Program. NASA. Online [available: *http://satc.gsfc.nasa.gov/support/ASM_FEB99/crm_at_nasa.html*]. Accessed 4/30/05.

NASA. 2004. Bioastronautics Critical Path Roadmap—An approach to risk reduction and management for human space flight: extending the boundaries. National Aeronautics and Space Administration. Rev 1 draft, 4/2/2004.

NASA. 2005. Bioastronautics Roadmap—a risk reduction strategy for human space exploration. On-line [available: *http://ston.jsc.nasa.gov/collections/TRS/-techrep/Sp-2005-6113.pdf*]. Accessed 1/6/2006.

NCRP (National Council on Radiation Protection and Measurements). 1996. A guide for uncertainty analysis in dose and risk assessments related to environmental contamination. NCRP Commentary No. 14. Bethesda, MD: NCRP.

NCRP. 1997. Uncertainties in fatal cancer risk estimates used in radiation protection. NCRP Report No. 126. Bethesda, MD: NCRP.

NCI (National Cancer Institute). 2003. Land CE, Gilbert E, Smith J, Hoffman FO, Apostoiae I, Thomas B, Kocher DM. *Report of the NCI–CDC Working Group to Revise the 1985 NIH Radioepidemiological Tables.* NIH Publication No. 03-5387. Bethesda, MD: U.S. Department of Health and Human Services, National Institutes of Health, National Cancer Institute.

Starr M, Chalmers I. 2005. The evolution of the Cochrane Library, 1988–2003. Update Software: Oxford. On-line [available: *www.update-software.com/history/clibhist.htm*]. Accessed 4/30/05.

Warren-Hicks WJ, Moore DRG (eds.). 1998. *Uncertainty Analysis in Ecological Risk Assessment.* Pensacola, FL: SETAC Press.

White RJ, Bassingthwaighte JB, Charles JB, Kushmerick MJ, Newman DJ. 2003. Issues of exploration: human health and wellbeing during a mission to Mars. *Adv. Space Res.* 31: 7–16.

Zadeh L. 1978. Fuzzy sets as a basis for a theory of possibility. *Fuzzy Sets and Systems* 1: 3–28.

4

Considerations Regarding the BR Context

The National Aeronautics and Space Administration's (NASA's) human space flight program has attained one of the world's greatest technological achievements: landing humans on the Moon. In 1966, at the height of the Apollo program, NASA received 4.4 percent of the total federal budget; today NASA receives less than 0.7 percent. Now NASA is poised to attempt missions of even greater visibility and risk. It is in this clash of performance expectations and budgetary constraints that the Bioastronautics Roadmap (BR) was created. This presents unique challenges (and opportunities) for NASA. This chapter considers the BR in the context of the pressures—internal and external to NASA—that generate additional risk for human space flight.

ORGANIZATIONAL CHARACTERISTICS AND RISK

The risks identified in the BR occur in the context of a larger set of risks to the human space flight program and to NASA as an organization. Highly visible failures, such as the loss of the Space Shuttles Challenger and Columbia, have the potential to erode public confidence in—and congressional support for—human space flight and NASA as an agency.

Although the presidential initiative announced in January 2004 (White House, 2004) has added impetus and focus to the goals of NASA, under certain circumstances it could add an additional risk—pressure being applied to achieve the goals of the initiative without sufficient time or resources for adequate preparation—that could compromise mission safety. Pressure could increase when critical biomedical research is delayed by a disaster-related response, such as the one that occurred after the loss of the Challenger. Thus, to the technical risks of space flight, the President's initiative has added the organizational risk that elements of the BR might be compromised in an effort to meet a societal goal. *The single most substantial organizational risk that NASA faces may be the possibility that a thoughtfully conceived roadmap could be preempted or abandoned as a result of such pressures or of an abrupt change in policy direction.*

Like the Challenger investigation, the Columbia Accident Investigation Board Report (CAIB, 2003) highlighted an inadequate safety culture within NASA leading to human performance failure. Figure 3.1 was included in this report because it illustrates the remarkable lack of agreement among a knowledgeable group of evaluators who were asked by NASA to address the question, How worried would you be about this risk if we were to go to Mars today? The responses were widely distributed, and mean or median values for such data would appear to be of little or no value. One of the key lessons from the Challenger and Columbia events was the importance of listening to even a single voice, if that voice came from a knowledgeable source, rather than responding to the "group mean" regarding risk.

These topics are conspicuously absent from the current version of the BR. Furthermore, differences in the organizational culture, and thus the safety culture, of the international space agencies participating in the International Space Station (ISS) or any of the future Design Reference Missions may exacerbate conflict both within crews and between crews and mission control, increasing the risk of human performance failure (NRC, 1998; Kanas et al., 2000). Support for this thesis is garnered from studies in analog environments, such as submarines (Wilken, 1969; Thomas et al., 2000) and Antarctic expeditions (Wood et al., 1999; Palinkas et al., 2004) that have noted cultural differences in interpersonal relations and adaptation to prolonged isolation and confinement as being relevant to BR Risks 24 and 25 (human performance failure due to poor psychosocial adaptation and human performance failure due to neurobehavioral problems) and, ultimately, to human performance failure.

There is a need to ensure close collaboration between NASA researchers, university- and foundation-based researchers, and operational personnel. Successfully implementing the BR will require working through or around this problem, bringing in various stakeholders (Palinkas et al.,

2005). The committee was especially sensitive to the relationships among internal NASA scientists, external investigators, and operations personnel, and these relationships were a prominent theme in many of the deliberations that led to the conclusions and recommendations in this report. Each of these communities feels, to some extent, that the other communities do not adequately appreciate their concerns or viewpoints, but this results in a creative tension that is highly appropriate because it brings advocacy to views that need to be represented in the risk analysis and mitigation approaches that make up the BR specifically, and the overall NASA bioastronautics program in general.

The committee concludes that these organizational and cultural factors can have important consequences for crew safety and mission success and thus represent risks that should be considered in the BR.

Recommendation 4.1
The committee recommends that an additional risk labeled "human performance failure due to organizational and cultural factors" be added to the BR. It may prove optimal to track this risk in a manner differently from the other risks in the BR (e.g., annual analyses of organizational and cultural risk in a separate report, use of an external standing panel to discuss this issue regularly). The committee's intent is that a risk-focused analysis of organizational and cultural issues become a visible part of the BR process.

ANALYSIS AND PRIORITIZATION TO MEET THE LAUNCH SCHEDULE

As a result of the President's space exploration initiative, NASA has proposed a schedule that requires considerable resources. Prospective funding, up-mass (determined by the type of launch vehicle), power, available equipment, and crew time (both number of crew and their availability to participate in bioastronautics research) are limited resources that directly affect NASA's ability to utilize the BR to reduce risk.

One example of constrained resources concerns the variety of countermeasures that are suggested for the inherent physiologic problems associated with exposure to the space environment for the period of time necessary to support the Design Reference Missions. Life support equipment that functions in microgravity for prolonged periods will have to be designed and tested, medical procedures that can be performed in a microgravity environment have to be created, and regenerative life support systems must be designed and built. For the above technologies, procedures, and capabilities to achieve a Technology Readiness Level (TRL) of 7, by definition these systems must be tested in an "operational environment" (i.e., a microgravity environment). It is almost axiomatic that the efficacy of alternative countermeasures—assuming the crew transits under microgravity conditions—can be tested only in an environment of microgravity. Therefore, the general problem of insufficient microgravity flight time, where such systems are tested and capabilities are validated for exploration class missions, becomes a formidable challenge. A similar concern exists for validation of systems and countermeasures in lunar and martian gravity.

Currently, various constraints—created by NASA or external to it, such as the Iran Nonproliferation Act of 2000—limit the International Space Station to a maximum of two (soon to be three) crew members. Routine maintenance of the ISS occupies the vast majority of a crew's time (NRC, 2002), leaving insufficient time for significant research activities, much less the effort that would be required to achieve a TRL of 7 for needed procedures. Furthermore, the committee is concerned that the planned ISS life span may not be sufficient to accommodate the necessary research or technology development and validation that will be necessary to enable the exploration vision. Finally, even if ISS support is extended and the crew size is augmented, it still may not fully meet the demands for the research that will be needed to support the Design Reference Missions. For example, without a test facility that closely duplicates the ambient pressure and partial pressures of gases to be found in the Crew Exploration Vehicle (CEV), appropriate decompression procedures for extravehicular activity (EVA) and space suit activities cannot be validated in an operational environment. Without a suitable low Earth orbit (LEO) research facility, it is not clear how NASA will be able to accomplish the research studies that are likely to be required to support the Mars initiatives.

Projects with extremely long lead times are particularly vulnerable to problems with research prioritization. Consider the derivation and validation of the select-in and select-out criteria for the crew selection process described in Chapter 2. In order to derive such selection criteria, long-duration isolation experiments under conditions of stress—using astronauts or astronaut surrogates as experimental subjects—would have to be performed during the derivation phase (Countermeasure Readiness Level, CRL, 1–3). A prospective validation phase would then be necessary (CRL 4–7). Such experiments would then have to be repeated to achieve a meaningful sample size (also requiring long-duration experiments with astronauts or their surrogates). It would be optimistic to believe that the first set of derived criteria would be successfully validated; therefore, several attempts at select-in and select-out criteria would be necessary (CRL 7–9).

Similarly, a decision to select a multicultural or mixed-gender crew for long-duration space flight appears likely. The evidence suggests that interpersonal dynamics are influenced substantially by factors such as cultural composition (Lozano and Wong, 1995; Kozerenko et al., 1999; Kring, 2001). Such team dynamics are important factors influencing the success of the expedition. Homogeneous teams appear to work better together than diverse teams (Chatman, 1991; Chatman et. al., 1998). If this experience is considered relevant to crew selection, the ISS or appropriate mission simulations would have to serve as test facilities to validate relevant select-in and select-out criteria for human planetary exploration.

It is not clear to the committee that select-in and select-out criteria could be successfully derived, validated, and implemented by the time human exploration beyond LEO commences.

Another example of a long-lead-time project concerns the question of whether an artificial gravity environment will be necessary to maintain crew health during a 30-month Mars mission. At issue is the fact that the 0.38*g* gravitational field of Mars cannot be simulated on Earth for more than a few seconds. Long-term simulations are possible using a centrifuge in the free-fall environment of the ISS. Such a facility—designed to fully support habitats for research rodents—is scheduled to launch in 2008 or later. However, this program is under consideration for cancellation as a result of budgetary constraints. Without this enabling research, it is entirely conceivable that astronauts would land on Mars without any evidence-based assurance that the martian gravitational field would provide sufficient musculoskeletal loading to ameliorate continued bone demineralization. Alternatively, large sums might be spent needlessly on developing a more complex rotating Mars spacecraft.

An ambitious and appropriate research program has been proposed by NASA in the BR. In view of the fact that the nation's most important test facility for human space flight research—the ISS—is constrained by time, up-mass, research facilities, escape capabilities, power, and crew availability

(NRC 2002; RAND, 2002), the committee concludes that resources are insufficient to perform the additional work necessary to mitigate the risks identified in the BR to acceptable levels.

Recommendation 4.2

The committee recommends that NASA perform regular, detailed assessments of the additional risks to the conduct of the President's 2004 vision for space exploration posed by the lack of available resources to fully address the issues posed in the BR. This assessment should then be used to make early strategic decisions regarding issues such as, but not limited to, the following:

1. How to provide support for a microgravity research platform that will have the resources (crew time, up-mass, facilities, and power) for the large amount of work necessary to validate countermeasures; achieve Technology Readiness Level 7 for life support systems sufficiently early in the design phase to allow their integration into the overall vehicle; and demonstrate the utility of medical procedures in microgravity.

2. How to support the extensive behavioral research program that would be necessary to validate processes or countermeasures such as select-in–select-out criteria (both for individual crew members and for a composite crew), issues related to cultural diversity, crew interactions, and isolation or stress-induced hazards. These issues may well require long lead times to study adequately.

ADDRESSING THE CHALLENGES POSED BY THE SMALL SAMPLE SIZE

A number of criteria can be considered in flight crew sizes: (1) resource requirements such as funding, vehicle capacity, and mission objectives; (2) standards for assessing quality control or hardware reliability; and (3) statistical power for performing research. Given the importance that NASA places on each of these criteria, the committee recognizes that NASA must consider all three sets of factors when determining crew sizes.

Regardless of which criteria are used to derive crew sizes, achieving statistically valid results using flight crews will be a daunting problem.[1] In general, very small sample sizes make it impossible to state either quality control or research findings with reasonable confidence intervals or to compare alternatives using tests of statistical significance. The committee recognizes that health-related studies based on observations of space mission crews will, for the foreseeable future, suffer from small sample size. Consequently, inferences based on single missions will have inadequate statistical power unless, in the context of reliability analysis, the problem under study is so prevalent that it is detected in the first few subjects (Virzi, 1992; Lewis, 1994). Methods are available to address this problem, including the pooling of data from multiple studies or missions in the manner of sequential clinical trials (IOM, 2001) and Bayesian sequential trials. *The committee proposes that rather than rely on data from a single mission for inference, NASA analyze data pooled from several missions.* More specifically, the committee proposes that studies be designed to incorporate as many missions as possible, somewhat in the manner of sequential clinical trials, and also that they incorporate prior information from archival data and ground-based studies to the extent practicable. In a Bayesian framework, a prior uncertainty distribution for extent of bone mass loss as a function of age, gender, and time in space, for example, would be incrementally modified by new information gained from—and incidental to—a series of missions. The goal would be to develop a sequence of posterior distributions about the

[1]By way of example, managers of NASA's bioastronautics program have tried to determine whether sufficient astronaut subjects will be available to conduct the forward research needed to successfully mitigate the risks imposed by the Design Reference Missions. A workshop was convened in Houston, Texas, in May 2004 to estimate the requirements for human subjects in exploration research. Representatives of the extramural and intramural research communities, flight surgeons, astronauts, bioastronautics management, and ISS payload planners attempted to derive realistic guesses of the minimum number of subjects that would be needed to mitigate each BR risk to an acceptable level. Without a priori calculation of statistical power, they concluded that at a minimum, 1,025 data points would be required to complete the logistical tasks associated with the current BR, with the majority (71%) derived from in-flight experiments. In the best case (i.e., if each ISS crew member participated in six experiments, the experiments did not interact, and they were performed without loss of data), at least 120 astronaut subjects would be required, nearly double the total number of ISS crew members expected for the rest of its useful life. (This assumes a three-person crew rotating every six months; this estimate and the utility of the resultant data would be further affected by changes in the size of the crew or the length of their sortie.)

quantity of interest, the last of which would always summarize the current accumulated information (see Appendix E for more details).

Recommendation 4.3

Drawing on the findings of the Institute of Medicine report *Small Clinical Trials: Issues and Challenges* (IOM, 2001), the committee recommends the use of pooled data from Bayesian sequential trials techniques and hierarchical random or fixed effects methods to compensate for the small sample sizes associated with individual flights.

EFFICIENCY AND TECHNOLOGY ISSUES

Bioastronautics is a focused effort to enable human exploration of space through effective risk management solutions and innovative science and technology discoveries (NASA, 2003). The BR states that the roadmap is the framework used to identify and assess the risks of *crew exposure to the hazardous environments of space* (NASA, 2005, p. 1). Later, risk is defined more broadly as *the conditional probability of an adverse event from exposure to the space flight environment* (NASA, 2005, p. 12). The lack of specific reference to crew exposure in this second definition has the potential to produce confusion and misinterpretation of the BR.

The BR states that it "guides the prioritized research and technology development that, coupled with operational space medicine, will inform: (1) the development of medical standards and policies; (2) the specification of requirements for the human system; and (3) the implementation of medical operations. The BR provides information that helps (1) establish tolerances (i.e. operating bands or exposure limits) for humans exposed to the effects of space travel and develop countermeasures to maintain crew health and function within those limits, and (2) develop technologies that make human space flight safe and productive."

In this context, it is inappropriate to have the BR address any other aspect of technology development that is not directly tied to *crew* exposure to the hazardous environments of space. The committee concludes that evaluating "efficiency risks" is relevant to the BR only when it relates specifically to crew exposure to the hazardous environments of space.

A more precise definition of "efficiency" may clarify the problem. Often, the BR uses the term "efficiency" ambiguously. Generically speaking, efficiency represents "the ratio of the effective or useful output to the total input in any system."[2] However, the efficiency risks described in the BR are described better by the term "economy,"[3] which represents savings realized by optimizing resource utilization. In the context of the BR—like many other space systems engineering problems—resources can be defined by six metrics: mass, volume, power, reliability (time), complexity, and consumables. These are legitimate system-level considerations, but the types of risks associated with thinking about these questions are quite different from those associated with crew exposures to hazards. The systems engineering approach required for exploration results in a solution that is economical in each of these dimensions. However, several of these dimensions are predefined for the human system. Thus, the risk management principles applied to such technology development will be different. For example, project risk management principles typically drive system engineering (Royer, 2002), and safety risk management principles typically drive the development of hardware and infrastructure in which and with which humans will interact.

The committee does not support the notion that risks associated with crew exposure to the hazardous environments of space and resource risks and constraints can credibly be addressed in parallel in the same management process. The proper time sequence is to address risks associated with crew exposure to the hazardous environments of space and continuously update the data available to a higher-level risk management process that makes more and more informed decisions about the value of accepting or further mitigating a risk. Resource constraints are appropriately addressed at this higher (programmatic) level, and these decisions feed back into overall systems designs, which are adjusted as needed to accommodate the total acceptable programmatic risk. This may well involve substantial engineering research and development to produce more effective or efficient system components, which is a key element of the spiral design concept that NASA has adopted for its exploration vision. However, the fundamental difference between the technology systems (e.g., vehicles; equipment; food preparation and delivery systems; clothing; air production, purification, conditioning, and distribution) and the human system is that the human system cannot be "reengineered" or redesigned in the same manner as mechanical components. In this sense, human risks drive all other systems development, with little room for adjustment.

---

[2] *http://dictionary.reference.com.*

[3] *http://dictionary.reference.com.*

The committee concludes that system efficiency concepts in the BR must focus on risks of adverse crew health events associated with technology and system failures.

Recommendation 4.4

The committee recommends that the current definition of risk be altered to clearly identify at least two types of risks: (1) health and medical risk, defined as the conditional probability of an adverse event to the human system (i.e., crew health or medical event) resulting from exposure to the space flight environment, and (2) engineering technology and system performance risk, defined as the conditional probability of an adverse event resulting from the space flight supersystem that affects crew health or mission success.

THE CASE OF ADVANCED HUMAN LIFE SUPPORT

The engineering and system technology risks found in the advanced human life support category are linked clearly to human health risks. Advanced human life support comprises food and life support systems, environmental monitoring and control systems, and EVA technologies and the human factors related to these technologies. In the area of Advanced Human Support Technologies, NASA faces challenges that may be divided into two areas: (1) determination of the optimal technology and (2) engineering development and qualification of the hardware, software, and operational procedures required to realize the system's performance. Neither of these challenges is associated directly with crew health risks, except through the development of medical and toxicological requirements that advanced human life support technologies must meet. Notwithstanding the recommendation above (i.e., that the BR should be focused only on engineering technology and system performance risks related to the conditional probability of an adverse crew health or medical event resulting from the space flight supersystem), the committee provides some discussion here of the two challenges.

Determining the optimal technology involves interrelated studies of the medical and toxicological, physical, chemical, and biological sciences and to date has built on accumulated experience. In the context of long-duration missions, ensuring highly reliable performance of technologies will depend on two principal means of verification: stress testing and full-duration life testing. In the former approach, relevant environmental factors are made more stressful (e.g., hotter or colder than normal) to permit evaluation of long-term performance in a short period of time. The "full-duration" approach is to build the apparatus and operate it within normal limits for an extended period of time, preferably several times the actual requirement. (As mentioned previously, in order to achieve a TRL of 7, this testing should be performed in a relevant environment.) Coupled with failure analysis and remediation, the full-duration approach gives the greatest confidence. To accomplish this sort of qualification with advanced life support systems, accumulated operational experience with such systems or their immediate predecessors is necessary.[4]

With regard to the underlying requirements for human health and performance that drive the operating parameters of advanced human life support systems, the committee believes that more attention must be paid to risks related to environmental factors associated with long-term missions, such as analyses of air and water quality, and factors related to crew cabin and EVA (e.g., suit, rover) atmospheric composition and pressure. Current operating parameters are derived from terrestrial standards and ground- and space-based operational testing environments available to date (including the ISS).

The committee found neither sufficient analysis of the research required to fully determine the operational characteristics for advanced life support system technologies for the Design Reference Missions nor evidence within the BR that justifies the use of terrestrial parameters for these missions.

REFERENCES

CAIB (Columbia Accident Investigation Board). 2003. Columbia Accident Investigation Board Report. On-line [available: *http://www.caib.us*]. Accessed 4/18/05.

Chatman JA. 1991. Matching people and organizations: selection and socialization in public accounting firms. *Administrative Science Quarterly* 36(3): 459–484.

[4]The Russian-built Elektron oxygen generator is a case in point. A U.S.-designed and built system using more advanced technology awaits launch in mid-2008. The United States is engaged in adapting the Russian system rather than using the intervening time to qualify the U.S. apparatus.

Chatman JA, Polzer JT, Barsade SG, Neale MA. 1998. Being different yet feeling similar: the influence of demographic composition and organizational culture on work processes and outcomes. *Administrative Science Quarterly* 43(4): 749–780.

IOM (Institute of Medicine). 2001. *Small Clinical Trials: Issues and Challenges.* Washington, DC: National Academy Press.

Kanas N, Salnitskiy V, Grund EM, Gushin V, Weiss DS, Kozerenko O, Sled AD, Marmar CR. 2000. Interpersonal and cultural issues involving crews and ground personnel during Shuttle/Mir space missions. *Aviat. Space Environ. Med.* 71(9 Suppl): A11–A16.

Kozerenko OP, Gushin VI, Sled AD, Efimov VA, Pystinnikova JM. 1999. Some problems of group interactions in prolonged space flights. *Hum. Perf. Extreme Environ.* 4(1): 123–127.

Kring JP. 2001. Multicultural factors for international spaceflight. *Hum. Perf. Extreme Environ.* 5(2): 11–32.

Lewis JR. 1994. Sample sizes for usability studies: additional considerations. *Human Factors* 36(2): 368–378.

Lozano ML, Wong CK. 1995. Human factors concerns for international partners in a Space Station environment. American Institute of Aeronautics and Astronautics Space Programs and Technologies Conference, Huntsville, AL, September 26-28. Herndon, VA: AIAA Publications Customer Service.

NASA (National Aeronautics and Space Administration). 2003. Bioastronautics Strategy. On-line [available: *http://spaceresearch.nasa.gov/docs/BioastronauticsStrategy.pdf*]. Accessed 5/18/05.

NASA. 2005. Bioastronautics Roadmap—a risk reduction strategy for human space exploration. On-line [available: *http://ston.jsc.nasa.gov/collections/TRS/-techrep/Sp-2005-6113.pdf*]. Accessed 1/6/2006.

NRC (National Research Council). 1998. *A Strategy for Research in Space Biology and Medicine into the Next Century.* Washington, DC: National Academy Press.

NRC. 2002. *Factors Affecting the Utilization of the Insternational Space Station for Research in the Biological and Physical Sciences (TGRISS Phase II).* Washington, DC: National Academy Press.

Palinkas LA, Glogower F, Dembert M, Hansen K, Smullen R. 2004. Incidence of psychiatric disorders after extended residence in Antarctica. *Int. J. Circumpolar Health* 63(2): 157–168.

Palinkas L, Lawrence A, Allred CA, Landsverk JA. 2005. Models of research—operational collaboration for behavioral health in space. *Aviat. Space Environ. Med.* 76 (1 Supp.): B52–B60.

RAND. 2002. *RAND Perspectives on ISS Budget Issues.* Arlington, VA: RAND Science and Technology Policy Institute.

Royer PS. 2002. *Project Risk Management Principles: A Proactive Approach.* Vienna, VA: Management Concepts, Inc.

Thomas TL, Hooper TI, Camarca M, Murray J, Sack D, Mole D, Spiro RT, Horn WG, Garland FC. 2000. A method for monitoring the health of U.S. Navy submarine crewmembers during periods of isolation. *Aviat. Space Environ. Med.* 71(7): 699–705.

Virzi RA. 1992. Refining the test phase of usability evaluation: How many subjects is enough? *Human Factors* 34(4): 457–468.

White House. 2004. President Bush announces new vision for space exploration program. Remarks by the President on U.S. space policy. On-line [available: *http://www.whitehouse.gov/news/releases/2004/01/20040114-3.html*]. Accessed 5/26/05.

Wilken DD. 1969. *Significant Medical Experiences Aboard Polaris Submarines: A Review of 360 Patrols During the Period 1963–1967.* Report No. 560. Washington, DC: Naval Submarine Medical Research Library.

Wood JA, Lugg DJ, Eksuzian DJ, Hysong SJ, Harm DL. 1999. Psychological changes in 100-day remote Antarctic field groups. *Environment and Behavior* 31: 299–337.

A

Bibliography of Related National Academies Reports

IOM (Institute of Medicine). 2001. *Safe Passage: Astronaut Care for Exploration Missions.* Washington, DC: National Academy Press.

IOM. 2001. *Small Clinical Trials.* Evans CH, Ilstad ST (eds.). Washington, DC: The National Academies Press.

IOM. 2004. *Review of NASA's Longitudinal Study of Astronaut Health.* Washington, DC: The National Academies Press.

NRC (National Research Council). 1983. *Risk Assessment in the Federal Government: Managing the Process.* Washington, DC: National Academy Press.

NRC. 1988. *Space Science in the Twenty-First Century: Imperatives for the Decades 1995 to 2015.* Washington, DC: National Academy Press.

NRC. 1996. *Radiation Hazards to Crews of Interplanetary Missions: Biological Issues and Research Strategies.* Washington, DC: National Academy Press.

NRC. 1997. *Advanced Technology for Human Support in Space.* Washington, DC: National Academy Press.

NRC. 1998. *A Strategy for Research in Space Biology and Medicine in the New Century.* Washington, DC: National Academy Press.

NRC. 2000. *Review of NASA's Biomedical Research Program.* Washington, DC: National Academy Press.

NRC. 2000. *Radiation and the International Space Station: Recommendations to Reduce Risk.* Washington, DC: National Academy Press.

NRC. 2003. *Factors Affecting the Utilization of the International Space Station for Research in the Biological and Physical Sciences.* Washington, DC: The National Academies Press.

NRC. 2004. *Issues and Opportunities Regarding the U.S. Space Program: A Summary Report of a Workshop on National Space Policy.* Washington, DC: The National Academies Press.

B

Methods

The committee developed this interim report and arrived at conclusions and recommendations regarding the strengths and weaknesses of the Bioastronautics Roadmap (BR) during a 13-month period from April 2004 to May 2005. During this time, the committee held four data-gathering sessions; made one site visit to the Johnson Space Center in Houston, Texas; and met in closed session seven times to deliberate. Agendas for the open, data-gathering sessions of these meetings are included in this appendix.

INSTITUTE OF MEDICINE
AGENDA

Committee on Aerospace Medicine and Medicine in Extreme Environments
and
Committee on Review of NASA's Bioastronautics Critical Path Roadmap
Keck Building
500 5th Street, NW
Washington, DC
Room 110

Monday, April 12, 2004

CLOSED SESSION
(committee and staff only)
8:00 a.m.–12:00 p.m.

OPEN SESSION

12:00 p.m. **Lunch**

1:00 p.m. **Request for a review of the Bioastronautics Critical Path Roadmap (BCPR)**
Richard Williams, M.D., Chief Health and Medical Officer, NASA

1:10 p.m. **Overview of the Bioastronautics Critical Path Roadmap**
NASA presenters: Guy Fogleman, director of bioastronautics research, Office of Biological and Physical Research; Howard Ross, acting deputy associate administrator for science, Office of Biological and Physical Research; Marc Shepanek, deputy chief, Medicine of Extreme Environments, NASA; Frank Sulzman, manager, Space Radiation Health Project

1:20 p.m. **How the Bioastronautics Critical Path Roadmap captures the critical risks and key research and technology issues for risk reduction, management, and informed decision making**

1:40 p.m.	**The Bioastronautics Critical Path Roadmap's method and expression of risk assessment for medical decision making and the BCPR communication of these methods for different mission scenarios**
2:20 p.m.	**Types of risks and impacts of risk as expressed in the Bioastronautics Critical Path Roadmap**
3:00 p.m.	**Break**
3:30 p.m.	**Categories of critical research issues and metrics used in the Bioastronautics Critical Path Roadmap**
4:10 p.m.	**Efficiency and technology issues in the Bioastronautics Critical Path Roadmap**
4:50 p.m.	**Plenary discussion** Led by *David Longnecker, M.D.*
5:30 p.m.	**Adjourn to reception and dinner with invited guests** *Location: third floor atrium*

Tuesday, April 13, 2004

OPEN SESSION

8:00 a.m.	**Continental breakfast**
8:30 a.m.	**Welcoming remarks** *David Longnecker, M.D.*
8:45 a.m.	**Overview of related work by the National Academies**
9:00 a.m.	**Space Studies Board/ASEB**
9:30 a.m.	**CAMMEE**
10:00 a.m.	**Break**

CLOSED SESSION
(committee and staff only)
10:15 a.m.–3:30 p.m.

INSTITUTE OF MEDICINE
AGENDA

Committee on Review of NASA's Bioastronautics Critical Path Roadmap
Keck Building
500 5th Street, NW
Washington, DC
Room 204

Wednesday, June 9, 2004

OPEN SESSION

9:00 a.m.	**Welcome and overview of day's agenda** *David Longnecker, M.D., and Lisa Vandemark, Ph.D.*
9:15 a.m.	**Briefings related to the review of the Bioastronautics Critical Path Roadmap** *NASA presenters: Lauren Leveton, Bioastronautics Science Management Team, NASA; Holly Patton, Aerospace Technologist, NASA; David Tomko, Lead Scientist, Biomedical Program, BASA Bioastronautics Research Division; Guy Fogleman, Director of Bioastronautics Research, Office of Biological and Physical Research; Frank Sulzman, Manager, Space Radiation Health Project*
12:00 p.m.	**Lunch**
1:00 p.m.	**Bone loss and countermeasures: historical perspectives and new in-flight clinical studies** *Jay Shapiro, M.D., Uniformed Services University of the Health Sciences*
2:00 p.m.	**Harmonization of crew living module and extravehicular pressure suit atmospheric constituents and pressures** *Bruce McCandless, M.S., M.B.A., Lockheed Martin*
3:00 p.m.	**Break**
3:30 p.m.	**An overview of space biology from cells to humans** *David Klaus, Ph.D., University of Colorado, Boulder*
4:30 p.m.	**Plenary discussion** Led by *David Longnecker, M.D.*
5:30 p.m.	**Adjourn**

Thursday, June 10, 2004

CLOSED SESSION
(committee and staff only)
8:30 a.m.–3:00 p.m.

INSTITUTE OF MEDICINE
AGENDA

Committee on Review of NASA's Bioastronautics
Critical Path Roadmap
Keck Building
500 5th Street, NW
Washington, DC
Room 201

Monday, August 2, 2004

OPEN SESSION

9:00 a.m.	**Welcome, introductions, and overview of day's agenda** *David Longnecker, M.D., and Lisa Vandemark, Ph.D.*
9:20 a.m.	**Briefings related to the review of the Bioastronautics Critical Path Roadmap** *Richard Williams, M.D.*
10:30 a.m.	**Break**
11:00 a.m.	**Briefings related to the review of the Bioastronautics Critical Path Roadmap** *NASA presenters via videoconference from Johnson Space Center: Guy Fogleman, Director of Bioastronautics Research, Office of Biological and Physical Research; Marc Shepanek, Deputy Chief, Medicine of Extreme Environments, NASA; Desmond Lugg, Chief, Medicine of Extreme Environments, Office of the Chief Medical Officer*
11:30 a.m.	**Question-and-answer discussion** *David Longnecker, M.D., Moderator*
12:00 p.m.	**Lunch**

1:00 p.m. **Advanced life support issues**
Brian Dunaway, Boeing Corporation

CLOSED SESSION
(committee and staff only)
2:30 p.m.–5:30 p.m.

Tuesday, August 3, 2004

CLOSED SESSION
(committee and staff only)
8:30 a.m.–3:00 p.m.

INSTITUTE OF MEDICINE
AGENDA

Committee on Review of NASA's Bioastronautics
Critical Path Roadmap
National Academy of Sciences Building
2101 Constitution Avenue, NW
Washington, DC
Lecture Room

Monday, January 31, 2005

OPEN SESSION

9:00 a.m. **Welcome and introductions**
David E. Longnecker, M.D., Chair

9:20 a.m. **Overview of the day's agenda**
Lisa M. Vandemark, Ph.D., Study Director

9:30 a.m. **Guest presentations**
Admiral Walter Cantrell, NASA Headquarters; Joseph Genovese, Hamilton Sunstrand; Joseph Fuller, Futron Corporation

12:00 p.m. **Lunch**

1:00 p.m.	**Guest presentations** *Geoffrey McIntyre, Federal Aviation Administration; Michael Gernhardt, NASA Johnson Space Center; Alan Feiveson, NASA Johnson Space Center (via phone conference); William Stone, Stone Aerospace*
3:15 p.m.	**Break**
4:00 p.m.	**Overview of the new Bioastronautics Roadmap** *Lauren Leveton, Ph.D., NASA Headquarters*
4:30 p.m.	**Plenary discussion** Led by *David Longnecker, M.D.*
5:30 p.m.	**Adjourn**

Tuesday, February 1, 2005

CLOSED SESSION
(committee and staff only)

C

Status of Countermeasure and Technology Readiness Levels

Appendix C lists the 183 projected deliverables from NASA's Bioastronautics Roadmap (BR). The developmental status for countermeasure readiness level (CRL) and technology readiness level (TRL) are shown for each deliverable. Dotted lines depict the range of CRL or TRL values when a range is projected. Deliverables marked as unknown were unclassified with respect to CRL & TRL at the time of BR release.

Risk	Title/Projected Countermeasures	CRL Ranking (black bar) / TRL Ranking (grey bar)									
			1	2	3	4	5	6	7	8	9
		Unknown	Definition	Preliminary Studies	Validated Hypothesis	Countermeasure Formulation	Proof of Concept	Clinical Testing	Space Flight Simulation	Flight Validation	Implementation
1	Bone Loss and Fracture Risk										
	• Biophysical modalities						x				
	• Crew screening		x								
	• Exercise and fitness regimens								x		
	• Hormone replacement therapy		x								
	• Nutrition					x					
	• Pharmacological								x		
	• Rehabilitation strategies				x						
	• Spacesuit design		x								
	• Artificial gravity	x									
2	Impaired Fracture Healing										
	• Biomechanical and pharmacological measures						x				
	• Ultrasound			x							
	• Minimize bone loss	x									
	• Development of treatment options	x									

3	Injury to Joints and Intervertebral Structures											
	• Improve fitness and conditioning											
4	Renal Stone Formation											
	• Nutrition											
	• Pharmacological agents											
	• Urine solubility testing in flight											
5	Occurrence of Cardiac Dysrhythmias											
	• Electrical cardioversion											
	• Nutritional countermeasure											
	• Pharmaceutical countermeasure											
	• Pre-flight and in-flight testing											
6	Diminished Cardiac and Vascular Function											
	• Artificial *g* exposure											
	• Drugs that affect cardiac mass function											
	• Improved exercise program											
7	Limits for Air and Water Contaminants											
	• Identify possible contaminants											

Risk	Title/Projected Countermeasures	CRL Ranking ■ / TRL Ranking ▒									
			1	2	3	4	5	6	7	8	9
		Unknown	Definition	Preliminary Studies	Validated Hypothesis	Countermeasure Formulation	Proof of Concept	Clinical Testing	Space Flight Simulation	Flight Validation	Implementation
8	Immune Dysfunction										
	• Definition of surrogate markers	x									
	• Prediction of crew susceptibility	x									
	• Detection systems			x							
9	Interaction Among Factors: Infections and Malignancy										
	• Antimicrobial agents					x					
	• Fusion proteins							x			
	• Molecular pathogen detection system								x		
	• Molecular diagnostic/detection kits							x			
	• Pathogen-specific immunizations			x							
	• Pre-flight microorganism screening			x							

10	Alterations in Microbes and Host Interaction										
	• Microbial identification technology						x				
	• Pre-flight screening								x		
	• In-flight microbial monitoring							x			
11	Reduced Muscle, Strength, and Endurance										
	• Artificial gravity				x						
	• Heavy resistance exercise							x			
	• Pharmacological interventions			x							
	• Biophysical interventions					x					
	• Improved endurance exercise							x			
	• Nutritional interventions							x			
12	Increased Susceptibility to Muscle Damage										
	• Artificial gravity				x						
	• Heavy resistance exercise							x			
	• Pharmacological interventions			x							

Risk	Title/Projected Countermeasures	CRL Ranking (solid bar) / TRL Ranking (shaded bar)									
			1	2	3	4	5	6	7	8	9
		Unknown	Definition	Preliminary Studies	Validated Hypothesis	Countermeasure Formulation	Proof of Concept	Clinical Testing	Space Flight Simulation	Flight Validation	Implementation
13	Impaired Sensory-Motor Capability During Flight, Entry, and Landing • Auto-land capability • Re-adaptation head movements • Workstation/spacecraft interior architecture design • Improve teleoperator displays • Gravity-specific pre-adaptation techniques • Pre-flight visual orientation for intravehicular activities training using VR techniques • Pre-flight training- fidelity simulator • Spatial ability tests • Evaluate in-flight landing rehearsals										

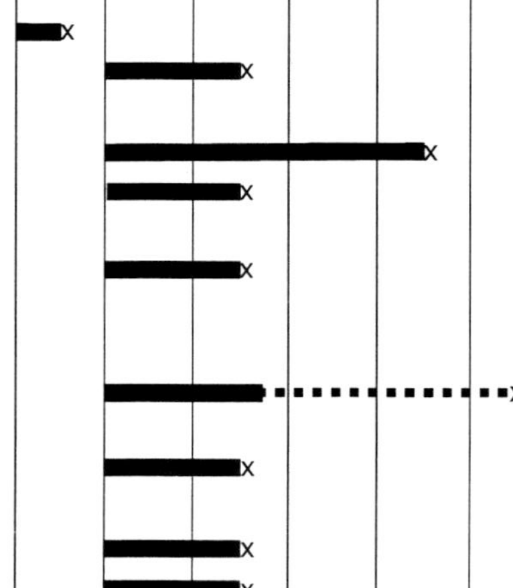

14	Impaired Sensory-Motor Capability After Landing • Balance prostheses • Gravity-specific pre-adaptation for landing • General *g*-specific pre-adaptation techniques (Mars) • Improved EVA suits • Pre-flight or in-flight *g*-specific pre-adaptation techniques (lunar) • Quantitative post-flight tests
15	Motion Sickness • New administration method for medications • Quantify cognitive deficits as a side effect of medication • Provide stroboscopic vision
16	Inadequate Nutrition • Improve dietary compliance • Enhance food system • Diet and nutritional supplementation • Refined nutritional requirements • Understanding and implementing an acceptable food system

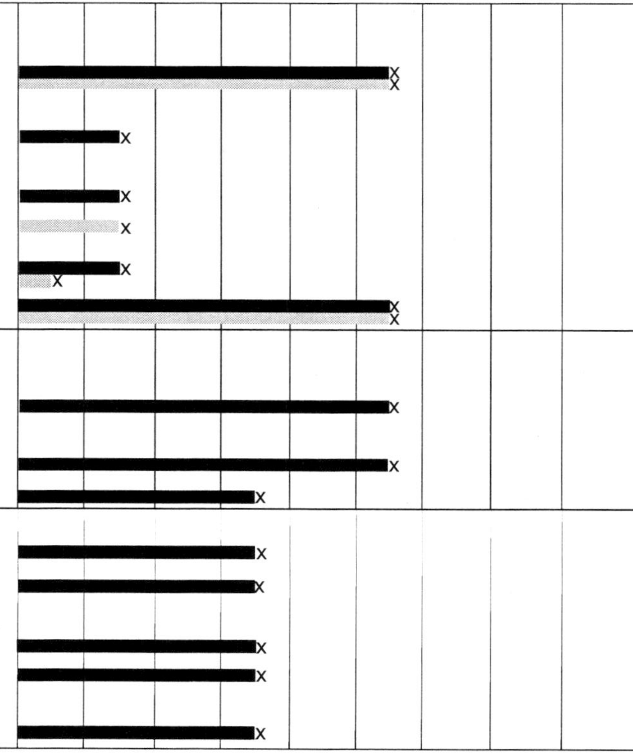

Risk	Title/Projected Countermeasures	CRL Ranking (black bar) / TRL Ranking (grey bar)									
			1	2	3	4	5	6	7	8	9
		Unknown	Definition	Preliminary Studies	Validated Hypothesis	Countermeasure Formulation	Proof of Concept	Clinical Testing	Space Flight Simulation	Flight Validation	Implementation
17	Monitory and Prevention										
	• Additional screening criteria	x									
	• Better in-flight health monitoring	x									
	• More autonomous and reliable diagnostics, care, and procedures	x									
18	Major Illness and Trauma										
	• Autonomous capabilities for monitoring and treatment	x									
19	Pharmacology of Space Medicine Delivery										
	• Shielding of medications	x									
	• Alterations in dose of medications	x									

20	Ambulatory Care										
	• More extensive medical kit	x									
	• More extensive telemedicine capabilities	x									
	• On-board medical diagnostics	x									
21	Rehabilitation on Mars										
	• Neurovestibular countermeasures	x									
	• Improved exercise protocols	x									
	• Autonomous medical monitoring	x									
	• Self-administered rehabilitation	x									
22	Medical Informatics, Technologies, and Support Systems										
	• Autonomous medical support system	x									
23	Medical Skill Training and Maintenance										
	• More extensive medical training	x									
	• Autonomous medical support	x									

Risk	Title/Projected Countermeasures	CRL Ranking ■ / TRL Ranking ░									
			1	2	3	4	5	6	7	8	9
		Unknown	Definition	Preliminary Studies	Validated Hypothesis	Countermeasure Formulation	Proof of Concept	Clinical Testing	Space Flight Simulation	Flight Validation	Implementation
24	Human Performance Failure due to Poor Psychosocial Adaptation										
	• Individual performance enhancement plans		x								
	• Individual and team selection				x						
	• Monitoring and early detection				x						
	• Predictive model of adaptability		x								
	• Select-in criteria	x									
25	Human Performance Failure due to Neurobehavioral Problems										
	• Greater observation by behavioral specialists during training					x					
	• Improved ability to manage uncooperative crewmember				x						
	• Improve remote diagnosis				x						
	• Improve cognitive self-assessment				x						

	• Individualized treatment algorithms						x				
	• On-board aids to manage stress and cognitive or emotional problems				x						
	• On-board modalities of therapy					x					
	• Technologies to detect stress and cognitive or emotional problems				x						
	• Predictive model				x						
	• Self-monitoring					x					
	• Updated behavioral medicine					x					
26	Mismatch Between Crew Cognitive Capabilities and Task Demands										
	• Design requirements for communications systems			x							
	• Onboard training systems			x							
	• Tools to analyze critical skills			x							
	• Tools to enable crew autonomy			x							
	• Tools to enable self-assessment			x							

Risk	Title/Projected Countermeasures	Unknown	1 Definition	2 Preliminary Studies	3 Validated Hypothesis	4 Countermeasure Formulation	5 Proof of Concept	6 Clinical Testing	7 Space Flight Simulation	8 Flight Validation	9 Implementation
	CRL Ranking ■ / TRL Ranking ░										
27	Circadian Rhythm Problems										
	• Ability to monitor sleep								x		
	• Flight rule limits on critical operations during sleep period					x					
	• Model sleep-related performance data							x			
	• Personal lighting device							x			
	• Non-photic adjustment tools						x				
	• Pharmacological interventions						x				
	• Photic adjustment tools								x		
28	Carcinogenesis										
	• Antioxidants		x								
	• Gene therapy		x								
	• Pharmaceuticals		x								
	• Improved shielding						x (TRL)				

29	Acute and Late Central Nervous System Risks										
	• Antioxidants		x								
	• Hydrogenous shielding						x				
	• Pharmaceuticals		x								
	• Autonomous monitoring	x									
	• Improved shielding materials	x									
	• Pharmacological cellular protectants	x									
30	Chronic and Degenerative Tissue Risks										
	• Antioxidants		x								
	• Hydrogenous shielding						x				
	• Pharmaceuticals		x								
	• Autonomous monitoring	x									
	• Improved shielding materials	x									
	• Pharmacological cellular protectants	x									
31	Acute Radiation Risk										
	• Antioxidants		x								
	• Hydrogenous shielding						x				
	• Pharmaceuticals		x								
	• Autonomous monitoring	x									
	• Improved shielding materials	x									
	• Pharmacological cellular protectants	x									

Risk	Title/Projected Countermeasures	CRL Ranking / TRL Ranking									
			1	2	3	4	5	6	7	8	9
		Unknown	Definition	Preliminary Studies	Validated Hypothesis	Countermeasure Formulation	Proof of Concept	Clinical Testing	Space flight Simulation	Flight Validation	Implementation
32	Monitor Air Quality										
	• Distributed network of detectors					x					
	• Highly sensitive analyzer suite					x					
33	Monitor External Environment										
	• Real-time radiation monitor					x					
	• Second-generation tear gas analyzer							x			
34	Monitor Water Quality										
	• On-line chemical water analyzer suite				x						
	• Microbial analysis instrument				x						

35	Monitor Surfaces, Food, and Soil										
	• Detection of surface contamination via optical interrogation				x						
	• Sampling methods			x							
36	Integrated Autonomous Control of Life Support Systems										
	• Automated control of life support			x							
37	Provide Space Suits and Portable Life Support Systems										
	• Cleaning/maintenance of soft goods	x									
	• Dust removal and prevention	x									
	• Increased in-orbit space suit life	x									
	• Longer shelf/service life of batteries	x									
	• Nonventing heat rejection system	x									
	• Reduced mass of suit	x									
	• Closed-loop CO_2 removal systems	x									

Risk	Title/Projected Countermeasures	Unknown	1 Definition	2 Preliminary Studies	3 Validated Hypothesis	4 Countermeasure Formulation	5 Proof of Concept	6 Clinical Testing	7 Space Flight Simulation	8 Flight Validation	9 Implementation
38	Maintain Food Quantity and Quality										
	• Assess psychosocial importance			x							
	• Determine effects of radiation		x								
	• Develop extended shelf life			x							
	• Increase variety and acceptability					x					
	• Hazard analysis					x					
	• High-barrier/low-mass packaging			x							
	• Refined nutritional requirements					x					
39	Maintain Acceptable Atmosphere										
	• Bioregenerative life support	x									
	• CO_2 moisture removal system					x					
	• Improved CO_2 removal/reduction					x					
	• Insitu resource utilization	x									
	• Trace contaminant control system					x					
	• Models to identify contaminant load	x									

CRL Ranking (black bar); TRL Ranking (gray bar)

40	Maintain Thermal Balance										
	• Thermal control system hardware							x			
41	Manage Waste										
	• Current practice might be adequate	x									
	• System for collecting waste			x							
	• System for transporting waste			x							
	• System for processing waste			x							
42	Bioregenerative Life Support Systems										
	• Integrated bioregenerative-PC testbed				x						
	• Low pressure greenhouse				x						
	• Mixed cropping systems						x				
	• Vegetable production unit						x				
	• Scale system to meet O_2/CO_2 requirements							x			
	• Scale salad production module to meet water and O_2 requirements					x					
43	Provide and Recover Potable Water										
	• Biological systems					x					
	• Possible in situ resource utilization	x									
	• Redundant systems			x							

Risk	Title/Projected Countermeasures	CRL Ranking / TRL Ranking									
			1	2	3	4	5	6	7	8	9
		Unknown	Definition	Preliminary studies	Validated hypothesis	Countermeasure formulation	Proof of concept	Clinical testing	Space flight simulation	Flight validation	Implementation
44	Mismatch Between Crew Physical Capabilities and Task Demands										
	• Measurement, analysis, modeling and design to optimize environment			x							
	• Tools to analyze physical tasks			x							
45	Poorly Integrated Ground, Crew, and Automation Functions										
	• Data on human performance			x							
	• Requirements for use of automated systems			x							
	• Tools for analyzing task requirements			x							

D

Bone Fracture Risk Associated with Prolonged Exposure to Microgravity

The incorporation of evidenced-based data into risk assessment is necessary to ensure that the Bioastronautics Roadmap (BR) evolves and remains current. Bone loss is a major risk that has been the focus of considerable research and countermeasure development at the National Aeronautics and Space Administration (NASA). However, the lack of evidence-based data for estimating the fracture risk associated with prolonged exposure to microgravity is a deficiency that compromises the assessment of countermeasure effectiveness. This is evident when one attempts to apply existing data to the following BR risk areas: (1) fracture risk assessment during prolonged microgravity exposure combined with a period of reduced-gravity exposure (e.g., during Mars exploration) and (2) fracture risk that is age-related due to early-onset osteoporosis as a consequence of bone loss following microgravity or reduced-gravity exposure.

Evidence-based data relating dual energy X-ray absorptiometry (DXA) bone density measurements to fracture risk do not exist for populations of men or women in the age range 35–50 years. Data do relate bone density measured by DXA to fracture risk, but for older, less physically fit populations. The World Health Organization (WHO) has defined "osteoporosis" as bone mineral density (BMD) greater than –2.5 standard deviations (SD) below age of peak bone density in white women (WHO, 1994). Fracture risk was estimated to increase 2.5-fold for each SD below adult peak bone density. However, data for men and non-white women were not included.

Melton et al. (1998) presented age-adjusted odds ratios for osteoporotic fracture per 1 SD decrease in BMD or in bone mineral apparent density (BMAD), by skeletal site, for individuals in Rochester, Minnesota. Men were 22–90 years of age (n = 348) and women were 21–98 years of age (n = 351). Per 1 SD decrease in BMD at total hip, for 29- to 49-year-old men, odds ratios were 1.38, 1.12, and 1.17 for total hip, femoral neck, and anterior–posterior (AP) spine scans, respectively, and 1.45 for total wrist. For women ages 20–49 years, odds ratios were 2.44, 1.72, and 1.59 for total hip, femoral neck, and AP spine scans, respectively, and 1.56 for total wrist. It is noted that unlike the astronaut crew, the baseline fracture rate was relatively increased in this community-based population. For men, total hip BMD was 1.061–0.140 g/cm^2, and for premenopausal women, mean total hip BMD was 0.941–0.124 g/cm^2. After adjusting for age, total hip BMD was the best predictor of fracture risk in women. In men, BMD at the wrist was the stronger predictor of fracture risk. This experience should be compared to existing data on baseline BMD and bone loss in the astronaut population.

It is widely appreciated that an assessment of fracture risk based solely on DXA BMD measurement is inadequate. Susceptibility to fracture at any level of bone mass is determined by bone density measurement and by the structural integrity of bone, which cannot be measured accurately with existing technology. The application of engineering principles to derive indirect assessments of bone strength—bone section modulus and buckling ratio (Beck et al., 2001; Kaptoge et al., 2003; Melton et al., 2005)—represents an advance in the field. Quantitative computed tomographic (QCT)–derived measurements of bone density and bone geometry in 14 astronauts after four to six months on the International Space Station (ISS) indicated a decline in bone density and parameters of bone strength (i.e., strength index, compressive strength indexes) in proximal hip and vertebral body (Lang et al., 2004). Consistent with earlier DXA studies involving Mir cosmonauts, bone was lost at rates of 0.8–0.9 percent per month at the spine and 1.2–1.5 percent per month at the hip. Although both cortical and trabecular bone declined, the percent loss in trabecular bone was greatest in the hip: proximal femur, 2.2–2.7 percent per month. As noted above, Lang et al. pointed out that the various bone strength indexes have not been validated as predictors of fracture risk and that the number of ISS subjects was small. Although presenting data were derived by a sensitive technique—volumetric QCT—it is likely that more accurate data relating bone loss to strength parameters will be derived in the near future from methods such as high-resolution anisotopic ultrasound or micromagnetic resonance imaging.

Thus, current estimates of fracture risk applied to space flight are inadequate for the following reasons:

- They are based on DXA data derived from older populations; these data could be applied to the older astronaut population but are not adequate for younger populations.
- Existing DXA data do not adequately address the problem of site specificity for bone strength, which is significant for estimating flight-related fracture risk.
- The contribution of prolonged residence in a reduced-gravity environment to either lessening bone loss or promoting bone formation immediately following extended microgravity exposure during transit cannot be assessed at this time.
- The added impact of a return flight from Mars on site-specific bone loss and fracture risk or the individual ability to regain bone mass at $1 \times g$ (as the astronaut ages) cannot be determined.

An evidenced-based assessment of bone loss and fracture risk (BR Risk 1, impaired fracture healing, and BR Risk 2, injury to joints and intervertebral discs) should be made using a multivariate analysis of existing flight-derived data on bone density and individual-specific characteristics that contribute to bone strength. The above risks are important during flight because of slow or poorly healing fractures in a microgravity environment and the need to anticipate these problems for medical care during flight. In addition, these risks are important because of back pain due to changes in intervertebral discs and the possibility of damage even after return from flight. As an example, factors for which evidence-based data can be obtained include the following:

- Age: susceptibility to fracture increases with increasing age.
- Prior history of fracture: prior fracture increases risk for future fracture.
- Family history of multiple fractures.
- Gender: men may fracture at greater BMD values than women (site specificity has not been determined, and the observation is limited in the numbers surveyed).
- Initial BMD: fracture rate increases as DXA BMD decreases.

• Return to baseline following successive flights as an indication of the resilience of bone-forming systems: failure to regain baseline BMD would presumably increase fracture risk when repeated microgravity exposure is experienced.

• Nutritional factors during and post-flight: maintenance of bone mass would be altered with prolonged negative intake of mineral and protein dietary components.

• Impact of exercise activity during and after flight: compliance with exercise and the effectiveness of specific exercise programs during and after flight impact bone mass.

• Rate and extent of bone loss on prior flights: there are no evidence-based data related to fracture risk for this factor.

• Alterations in structural and geometric parameters that could have iterative effects on bone strength occurring during successive flights.

• Duration of microgravity or reduced-gravity exposure: it has been demonstrated that six-month microgravity exposure reduces bone mass in most astronauts; however major fractures have not been reported, nor has the occurrence of stress fractures following return to 1*g*.

• Fracture risk assessment during and following extended microgravity exposure requires the utilization of existing astronaut data as well as the expansion of current methods for measuring bone mass and bone strength.

The dearth of data related to the microgravity environment encountered during extended-duration space flight suggests a need for increased research to support evidence-based decision making regarding these subjects.

REFERENCES

Beck TJ, Oreskovic TL, Stone KL, Ruff CB, Ensrud K, Nevitt MC, Genant HK, Cummings SR. 2001. Structural adaptation to changing skeletal load in the progression toward hip fragility: the study of osteoporotic fractures. *J. Bone Min. Res.* 16(6): 1108–1119.

Kaptoge S, Dalzell N, Loveridge N, Beck TJ, Khaw KT, Reeve J. 2003. Effects of gender, anthromorphic variables and aging on the evolution of hip strength in men and women over 65. *Bone* 32: 561–570.

Lang T, LeBlanc A, Evans H, Ying L, Genant H, Yu A. 2004. Cortical and trabecular bone mineral loss from the spine and hip in long-duration space flight. *J. Bone Min. Res.* 19(6): 1006–1012.

Melton LJ III, Atkinson EJ, O'Connor MK, O'Fallon WM, Riggs BL. 1998. Bone density and fracture risk in men. *J. Bone Min. Res.* 13(12): 1915–1923.

Melton LJ III, Beck TJ, Amin S, Khosla S, Achenbach SJ, Oberg AL, Riggs BL. 2005. Contributions of bone density and structure to fracture risk assessment in men and women. *Osteoporos. Int.* Feb. 2. Abstract on-line [available: *http://www.ncbi.nlm.nih.gov/entrez/query.fcgi?cmd=Retrieve&db=pubmed&dopt=Abstract&list_uids=15688123*]. Accessed 4/15/05.

WHO (World Health Organization). 1994. *Assessment of Fracture Risk and Its Application in Screening for Postmenopausal Osteoporosis.* Technical Report Series 843. Geneva: World Health Organization.

E
Integration of Data-Based Evidence and Expert Opinion in Decision Making

BAYESIAN UPDATING

In certain cases, it may be necessary to update research data with new findings and with stakeholder opinions (where stakeholders are defined to be mission specialists, National Aeronautics and Space Administration [NASA] directors, managers, and flight surgeons). Bayesian updating may be one strategy for integrating stakeholder opinions with data from research studies. Such a strategy can accommodate contrasting points of view from stakeholders expressed in a subjective manner. It is necessary that the data from different research studies measure the same underlying factor (e.g., diastolic blood pressure, depression) on the same scale (e.g., millimeters of mercury for blood pressure, Hamilton scale for depression).

This updating strategy consists of the following steps: (1) selection of a sample of stakeholders; (2) elicitation of probability information from these stakeholders; (3) translation of this information to statistical distributions, called "prior" distributions, for each contrasting view of the stakeholders; (4) assignment of an "importance" weight to each of these prior distributions for each of the contrasting views; (5) with these importance weights, derivation of a "summary prior" distribution by taking a weighted combination of the contrasting prior distributions; (6) derivation of a "summary likelihood" pooling all study datasets while accounting for the varying variability and sample sizes across the study datasets; (7) derivation of a "summary posterior" distribution from the summary prior distribution and summary likelihood; (8) choosing a utility function to incorporate costs and stakeholders' sensitivity to such costs; and (9) decisions based on regrets or opportunity costs in cost–benefit or risk–benefit models by weighing outcome information from summary posterior distribution (e.g., mean differences, risk differences, risk ratios, odds ratios, and interactions involving these effects) against utility functions. Each of these steps is described in more detail below.

- *Steps 1–2. Selection of a sample of stakeholders and elicitation of probability information from these stakeholders:* The selection process should at least be comprehensive, maximizing the number of contrasting points of view, if the process is not random. Stakeholders' prior opinions will be elicited with questionnaires. In these questionnaires, stakeholders will be asked to provide ranges of probabilities of confidence in positive and negative results. The design of the questionnaire will be selected from several different designs available in the research literature. Chaloner and Rhame (2001) presented an interactive approach based on iterative elicitation from physicians enhanced by real-time iterative and graphical feedback to the physicians of their quantified opinions. Parmar et al. (1994, 2001) and Spiegelhalter et al. (1994) presented a questionnaire for eliciting prior distributions in a pair of large randomized trials conducted as part of the British Medical Research Council Cancer Trials. In a hepatocellular carcinoma clinical trial, Tan et al. (2003) also used such a questionnaire to elicit prior information on the equivalence between surgery with adjuvant therapy versus surgery alone on recurrence-free survival. Alternatively, a series of individual description formats have been developed by Vennix et al. (1994). These individual questionnaire formats focus on the different phases proposed by for elicitation: (1) the positioning phase, which defines the context of the information; (2) the description phase, which guides stakeholders through four aspects of description (visual, verbal, textual, and graphic); and (3) the discussion phase, in which the individual descriptions from phase 2 are examined and compared.
- *Step 3. Translation of this information to prior distributions:* Individual histograms representing the prior beliefs of each investigator can be constructed from the relative probability values that stakeholders may be asked to provide in Step 2. Following Spiegelhalter et al. (1994), these probability values may be summarized across stakeholders with similar opinions to then construct "overall histograms"and "skeptical histograms."

These histograms will represent the overall (or clinical) and skeptical (or cautious) prior distributions associated with the stakeholders' views of the problem. A skeptical prior distribution corresponds to the beliefs of individuals who are reluctant to accept alternative hypotheses of interest to the investigators. The resulting histograms can be transformed to the scale on which research data have been collected (Tan et al., 2003).

- *Steps 4–7. Determination of multivariate prior distributions from multiple stakeholders, estimation of summary likelihoods from multiple datasets, resulting in a derivation of a summary posterior distribution:* Such a procedure is based on Bayes' rule and entails intractable integration resulting in simulation-based integration (e.g., Markov Chain Monte Carlo), which many commercially available software packages now offer (Spiegelhalter et al., 1994).
- *Steps 8. Choosing a utility function to incorporate costs and stakeholders' sensitivity to such costs:* Such functions involve determining costs from implementing mitigation strategies and reduced costs from preventing problem outcomes (Pliskin et al., 1980; Berger, 1985; Lindley, 1985; Gold et al., 1996). Sensitivity analysis of the overall procedure outlined here includes varying such cost estimates (Matchar and Samsa, 1999; Matchar et al., 1997).
- *Step 9. Decisions based on regrets or opportunity costs from weighing information on outcomes under mitigation strategies against outcomes under the absence of mitigation strategies:* Regrets are based on loss functions as contrasts between decisions that lead to optimal utility benefits and the utility benefits based on observed or predicted data. The expected loss functions allow the incorporation of research data and previous opinions of stakeholders by integrating utility functions for the optimal and observed decisions with respect to the summary posterior distributions (Berger, 1985: Lindley, 1985). For complex sequences of branching decisions based on outcomes of previous decisions, backward induction algorithms may be used (Bellman, 1957).

Overall, such decision processes present a complex web of different statistical procedures, research datasets, and opinions by stakeholders. This complex web is sensitive to selected procedures and corresponding assumptions; thus, this sensitivity is assessed by varying assumptions and operational procedures (Matchar and Samsa, 1999). Varying assumptions, procedures, and information used for forming utility functions and prior distribution may be done formally with quantified ranges of possible values and model-averaging techniques or informally by choosing plausible values of model and prior distribution parameters (Spiegelhalter et al., 1994).

REFERENCES

Bellman RE. 1957. *Dynamic Programming.* Princeton, NJ: Princeton University Press.

Berger JO. 1985. *Statistical Decision Theory and Bayesian Analysis,* 2nd Ed. New York: Springer-Verlag.

Chaloner K, Rhame FS. 2001. Quantifying and documenting prior beliefs in clinical trials. *Statistics in Medicine* 20: 581–600.

Gold MR, Siegel JE, Russell LB, Weinstein MC (eds). 1996. *Cost-Effectiveness in Health and Medicine.* Oxford, UK: Oxford University Press.

Lindley DV. 1985. *Making Decisions,* 2nd Ed. New York: Wiley.

Matchar DB, Samsa GP. 1999. Using outcome data to identify best medical practice: the role of policy models. *Hepatology* 29 (6 Suppl): 36S–39S.

Matchar DB, Samsa GP, Mathews JR, Aneukiewicz M, Parmigiani G, Hasselblad V, Wolf PA, D'Agostino RB, Lipscomb J. 1997. The Stroke Prevention Policy Model (SPPM): linking evidence and clinical decisions. *Ann. Int. Med.* 127: 704–711.

Parmar MKB, Griffiths GO, Spiegelhalter DJ, Souhami RL, Altman DG, van der Scheuren E, CHART steering committee. 2001. Monitoring of large randomised clinical trials: a new approach with Bayesian methods. *Lancet* 358(9279): 375–381.

Parmar MKB, Spiegelhalter DJ, Freedman LS. 1994. The CHART trials: Bayesian design and monitoring in practice. *Statistics in Medicine* 13: 1297–1312.

Pliskin JS, Shepard D, Weinstein MC. 1980. Utility functions for life years and health status: theory, assessment, and application. *Operations Research* 28: 206–224.

Spiegelhalter DJ, Freedman LS, Parmar MKB. 1994. Bayesian approaches to randomised trials. *J. R. Stat. Soc. A* 157: 357–387.

Tan SB, Chung YF, Tai BC, Cheung YB, Machin D. 2003. Elicitation of prior distributions for a phase III randomized controlled trial of adjuvant therapy with surgery for hepatocellular carcinoma. *Control. Clin. Trials* 24(2): 110–121.

Vennix JAM, Anderson DF, Richardson GP, Rohrbaugh J. 1994. Model building for group decision support: issues and alternatives in knowledge elicitation. In Morecroft J, Sterman J. (eds.) *Modeling for Learning Organizations.* Portland, OR: Productivity Press.

F

Biographical Sketches of Committee Members and Staff

COMMITTEE

DAVID E. LONGNECKER, M.D. (*Chair*), is the Robert Dunning Dripps Professor Emeritus of Anesthesiology and Critical Care at the University of Pennsylvania and a director in the Division of Health Care Affairs at the Association of American Medical Colleges (AAMC), Washington, DC. Dr. Longnecker received his M.D. degree from Indiana University School of Medicine, where he completed residency training in anesthesiology. Following an NIH Special Research Fellowship in cardiovascular physiology at Indiana U., he continued clinical and laboratory research at the NIH Clinical Center, where he served as a clinical associate. Thereafter, he was an assistant professor in the Departments of Anesthesiology and Physiology at the University of Missouri and then the Harold Carron Professor of Anesthesiology at the University of Virginia. He has received numerous NIH research grants and a Research Career Development Award for research involving the effects of anesthetics on the microcirculation, oxygen delivery to tissue, oxygen therapeutics, endothelium-dependent circulatory control, and health services research. Dr. Longnecker is the author or co-author of more than 200 scientific abstracts, original scientific articles, and book chapters and the editor of five textbooks of anesthesiology. He is a member (by election) of the Royal College of Anaesthetists (UK) and the Institute of Medicine. He chairs the IOM Committee on Aerospace Medicine and Medicine for Extreme Environments and previously chaired IOM committees on Fluid Resuscitation for Combat Casualties and on the Longitudinal Study of Astronaut Health.

JAMES P. BAGIAN, M.D., P.E., is the director, Department of Veterans Affairs National Center for Patient Safety. From 1980 to 1995, Dr. Bagian served as a NASA astronaut. He is a veteran of two shuttle missions, including the first dedicated Space and Life Sciences Spacelab mission. He was also a lead investigator for both the Challenger and Columbia accidents. Dr. Bagian focuses on applications in aerospace systems, notably crew survival and physiological adaptation issues that impact aviation and space flight operations, as well as environmental technology. He has also developed and implemented, on national and international bases, systems-based solutions to improve patient safety. Dr. Bagian is a member of IOM and NAE and has served on or chaired numerous committees of the National Academies.

ELIZABETH R. CANTWELL, Ph.D., is the deputy division leader for science and technology in the International, Space and Response Division at Los Alamos National Laboratory. Until June 2005, she served as the section leader for the Micro and Nanotechnology Center, Lawrence Livermore's engineering research center for fabricating small sensors and devices. She earned an undergraduate degree in psychology from the University of Chicago and an M.S. and a Ph.D. in mechanical engineering from the University of California at Berkeley. She also holds an MBA from the University of Pennsylvania's Wharton School of Business. Dr. Cantwell began her career building life support systems for manned space missions with the National Aeronautics and Space Administration (NASA) and was a member of the NRC Committee on Advanced Technology for Human Support in Space (1996–1997). She is currently a member of the IOM Committee on NASA's Bioastronautics Critical Path Roadmap.

VALERIE GAWRON, Ph.D., is a technology fellow at General Dynamics in Buffalo, New York. Dr. Gawron received a Ph.D. in engineering psychology from the University of Illinois and master's degrees in experimental psychology, industrial engineering, and business administration from the State University of New York. She is a fellow of the Human Factors and Ergonomics Society and an associate fellow of the American Institute of Aeronautics and Astronautics, with previous NRC service. She was a member of the Air Force Scientific Advisory Board and is now a member of the Army Science Board. Currently, her research focuses on the cognitive and environmental effects of human performance, with a specialization in situational awareness, workload testing, and evaluation. She is also the chair of the Science and Technology Working Group of NASA's Space–Human Factors Engineering Group.

CHRISTOPHER A. HART, J.D., is assistant administrator for system safety at the FAA. Mr. Hart holds a B.A. and an M.A. in aerospace and mechanical science from Princeton University. He also earned his Juris Doctor from Harvard Law School. He holds a commercial pilot's license with multi-engine and instrument ratings as well. Mr. Hart served as a member of the National Transportation Safety Board (1990–1993), where he had specialized interests in human factors and the impact of automation on transportation systems. He was nominated for this committee because of his expertise in the technical aspects of risk assessment and decision making, and his familiarity with aerospace technology.

CHARLES E. LAND, Ph.D., is a senior investigator with the National Cancer Institute (NCI). He received his Ph.D. in statistics from the University of Chicago, studied risk of radiation-related cancer at the Atomic Bomb Casualty Commission and the Radiation Effects Research Foundation in Hiroshima, Japan, and taught statistics at Oregon State University before joining the NCI in 1975. His interests center on quantification of radiation-related cancer risk based on epidemiological studies of exposed populations, the role of uncertain assumptions needed to obtain such estimates, and the public policy implications of uncertainty in estimated risk, especially as it bears on risks at low doses. Dr. Land has served on expert committees of the National Academy of Sciences, the National Council on Radiological Protection and Measurements, the International Commission on Radiological Protection, and other organizations. He is a fellow of the American Statistical Association. Dr. Land was nominated for this committee because of his expertise in the statistical analysis of cancer risk from radiation exposure.

DANIEL R. MASYS, M.D., is professor and chair of the Department of Biomedical Informatics, and professor of medicine at the Vanderbilt University School of Medicine. Previously he served as director of biomedical informatics and professor of medicine at the University of California, San

Diego School of Medicine. An honors graduate of Princeton University and the Ohio State University College of Medicine, he completed postgraduate training in internal medicine, hematology, and medical oncology at the University of California, San Diego and the Naval Regional Medical Center, San Diego. Previously, he served as chief of the International Cancer Research Data Bank of the National Cancer Institute, National Institutes of Health, and from 1986 through 1994 was director of the Lister Hill National Center for Biomedical Communications. In this capacity, Dr. Masys served as the chief program architect and first director of the National Center for Biotechnology Information (NCBI) that was established within the National Library of Medicine in 1987 to support molecular databases and computational tools. Dr. Masys is a diplomate of the American Board of Internal Medicine in medicine, hematology, and medical oncology. He is a fellow of the American College of Physicians, a fellow of the American College of Medical Informatics, and a member of the Institute of Medicine. He has served as a consultant to the NASA Life Science Informatics program and is an active instrument-rated pilot.

BRUCE M. MCCANDLESS II, M.B.A., is a veteran astronaut, having served in that capacity for 24 years, and made two flights in the Space Shuttle. He performed two EVAs or "space walks," making the first solo flights in the Manned Maneuvering Unit. He is an experienced SCUBA diver and has been involved in numerous hypobaric activities within NASA. He is currently an aerospace engineer and Principal Research Scientist within the Civil Space product area of Lockheed Martin Space Systems Company in Denver, Colorado. Mr. McCandless has directed several space technology risk assessment efforts, including the first phase of the Jupiter Icy Moons Orbiter nuclear-fission powered spacecraft studies. He was nominated for appointment to this committee because of his wide range of expertise in analyzing and managing risk associated with failure of human support technology systems in space.

TOM S. NEUMAN, M.D., is professor of medicine and surgery, director of the Hyperbaric Medicine Center, associate director of the Department of Emergency Medicine, and attending physician, Emergency Department at the University of California, San Diego Medical Center. A graduate of Cornell University, he received his M.D. from the New York University School of Medicine in 1971, followed by internship and residency in internal medicine at Bellevue Hospital. Dr. Neuman is board certified in internal medicine, pulmonary disease, occupational medicine, undersea and hyperbaric medicine, and emergency medicine. He is a fellow of the American College of Physicians and the American College of Preventive Medicine. Dr. Neuman has been a leader in the field of the physiology and medicine of diving throughout his career. He previously served on two IOM committees, which published the following reports: *Review of NASA's Longitudinal Study of Astronaut Health* and *Safe Passage: Astronaut Care for Exploration Missions*. He was nominated for this committee for his knowledge of undersea and hyperbaric medicine and related occupation health risks.

THOMAS F. OLTMANNS, Ph.D., is the Edgar James Swift Professor of Arts and Sciences in the Department of Psychology at Washington University in St. Louis, Missouri. He previously served as professor of psychology and psychiatric medicine and director of clinical training in psychology at the University of Virginia. He has also served as professor of psychology at Indiana University. Dr. Oltmanns received his undergraduate degree from the University of Wisconsin–Madison and his Ph.D. at the State University of New York at Stony Brook. He has authored 5 books and more than 70 journal articles. Dr. Oltmanns is past president of the Society for a Science of Clinical Psychology and is a consulting editor for the *Journal of Abnormal Psychology* and a member of the editorial boards of *Psychological Bulletin* and *Journal of Personality Disorders*. His research has been supported by numerous grants, and he is currently co-principal investigator on a large grant looking at peer assessment of personality traits and pathology. He has served on two different grant review committees for the National Institute of Mental Health and is a member of NASA's Astronaut Selection Psychiatric Standards Working Group.

LAWRENCE A. PALINKAS, Ph.D., is a professor in the School of Social Work at the University of Southern California. Dr. Palinkas serves as the deputy chief officer of the Life Sciences Standing Scientific Committee of the Scientific Committee on Antarctic Research (SCAR). He has more than 15 years of experience in studying behavioral adaptation in the Antarctic. He has also been active in translating Antarctic research for use in developing effective countermeasures to long-duration missions in space. Dr. Palinkas served as a member of the National Academy of Science's Committee on Space Biology and Medicine from 1997–2000. In 1997–1998, he served as chair of the Behavior and Performance Panel and principal author of the chapter on Behavior and Performance in the committee's 1998 report *A Strategy for Research in Space Biology and Medicine in the New Century* (National Academy Press). He also reviewed NASA's current research efforts in behavior and performance in the committee's 2000 report *Review of NASA's Biomedical Research Program* (National Academy Press). He currently serves as chair of the External Advisory Council of the National Space Biomedical Research Institute (NSBRI) and as a member of the Behavior and Performance Integrated Product Team at NASA's Johnson Space Center.

JAMES PAWELCZYK, Ph.D., is an associate professor of kinesiology, physiology, and medicine at the Pennsylvania State University. He was a payload specialist on the STS-90 (Neurolab) mission, which flew in 1998 with a focus on neuroscience. Dr. Pawelczyk was a member of NASA's Life Sciences Advisory Subcommittee, Office of Biological and Physical Research, 1998–2002, and a member of ReMaP Task Force, 2002, which was charged with establishing priorities for research on the International Space Station. He has received NASA funding as an individual principal investigator and a project leader on center grants and contracts (including international collaboration) since 1993. Dr. Pawelczyk's research areas include central neural control of the cardiovascular system and compensatory mechanisms to conditioning and deconditioning. He was nominated for this committee because of his familiarity with NASA and space flight, as well as for his medical expertise in the effects of space travel on human systems.

BRUCE S. RABIN, M.D., Ph.D., is a professor of pathology and psychiatry at the University of Pittsburgh Medical Center, medical director of the Clinical Immunopathology Laboratory, and medical director of the Healthy Lifestyle Program. His research areas of interest and capability are in the interrelationship between stress, immune function, and health. These are critical to several concerns of the Bioastronautics Critical Roadmap. Dr. Rabin is past president of the Psychoneuroimmunology Research Society. He has served on a number of government panels to promote research in mind–body interactions. Dr. Rabin was nominated for this committee because of his interdisciplinary research into the effects of stress on human body systems, including several disciplines germane to this study such as immunology and human behavior changes.

KARLENE ROBERTS, Ph.D., is a professor in the Haas School of Business of the University of California, Berkeley and a research psychologist at Berkeley's Institute of Industrial Relations. Dr. Roberts has expertise in the design and management of organizations and systems of organizations in which errors can have catastrophic consequences. The results of her research have been applied to programs in numerous organizations including the U.S. Navy and Coast Guard, the Federal Aviation's Air Traffic Control System, NASA, and the medical industry. Dr. Roberts has published on a wide range of organizational risk management issues. She is a fellow in the American Psychological Association, the American Psychological Society, and the Academy of Management. Dr. Roberts was nominated for this committee because of her expertise in human psychology and behavior and organization risk management strategies.

CAROL E. H. SCOTT-CONNER, M.D., Ph.D., is professor, Department of Surgery, University of Iowa, Iowa City. Dr. Scott-Conner received her M.D. from the New York University School of Medicine in 1976 and stayed at NYU for her surgical residency, which she completed in 1981. After leaving NYU, she joined the faculty at Marshall University and then moved to the University of Mississippi. During her tenure there she earned a Ph.D. in anatomy from the University of Kentucky and an M.B.A. Since 1995, she has been professor and head of surgery at the University of Iowa. Dr. Scott-Conner has been active on 22 editorial boards, and has authored more than 200 original papers, abstracts, reviews, and book chapters. She holds memberships in many elected surgical societies and has frequently served in leadership positions. She previously served as a member of the IOM Committee on Space Medicine. Dr. Scott-Conner was nominated to this committee because of her broad expertise in clinical care related to astronaut health and risk issues.

RHEA SEDDON, M.D., is the assistant chief medical officer for the Vanderbilt Medical Group in Nashville, Tennessee, where she has worked extensively on patient safety and quality improvement. A former three-flight veteran astronaut for NASA, she logged more than 722 hours in space. She was a mission specialist on STS-51D and STS-40 and was the payload commander on STS-58. Dr. Seddon also served in several other capacities at NASA, namely as technical assistant to the director of flight crew operations and special adviser for Shuttle/Mir scientific payloads and as a member of NASA's Aerospace Medical Advisory Committee and the International Bioethics Task Force. After earning a B.A. in physiology at the University of California at Berkeley and an M.D. from the University of Tennessee, Dr. Seddon went on to complete an internship and residency in general surgery in Memphis. She was nominated to this committee because of her familiarity with NASA and manned space flight and her clinical background in patient care.

JAY R. SHAPIRO, M.D., professor, Department of Physical Medicine and Rehabilitation, Johns Hopkins University, formerly served as director of the Interdepartmental Center for Space Medicine, Uniformed Services University and is currently director of the Osteogenesis Imperfecta Clinic at the Kennedy-Krieger Institute. Dr. Shapiro also served as team lead for bone, National Space Biomedical Research Institute. Dr. Shapiro has many years of direct experience with NASA research and clinical countermeasures related to bone and muscle loss in a microgravity environment. He is nominated for this committee because of his historical perspective on NASA risk management of bone loss and his expertise in a wide range of clinical countermeasures, including the application of in-flight drug trials related to bone loss.

THOMAS TEN HAVE, Ph.D., is professor of biostatistics with training in biostatistics at the University of Michigan (B.A. in statistics, M.P.H. and Ph.D. in biostatistics). He has statistical research interests in categorical data analysis, random effects models, informative dropout, treatment non-adherence, and designs and statistical analyses to accommodate patient preferences and adaptive treatment regimes. This methods research melds with his collaborations in psychiatry, family medicine, addiction research, and disparities research, with a focus on multi-site randomized and observational studies. Dr. Ten Have's research is facilitated by his roles as the principal investigator of National Institute of Mental Health (NIMH)-funded R01 and T32 training grants and the director of the Biostatistics-Data Core and co-investigator of the NIMH-sponsored Advanced Center for Intervention Services Research (ACISR) for Depression in the Aged. Finally, Dr. Ten Have is strongly committed to affirmative action in the recruitment of students, faculty members, investigators, study participants, and research topics. Among his contributions to the profession, Dr. Ten Have is a fellow of the American Statistical Association and associate editor of *Biometrics.*

IOM COMMITTEE STAFF

RICARDO A. MOLINS, Ph.D., is a senior program officer. Since joining the IOM in 1999, he has been the study director of the Food Chemicals Codex with the Food and Nutrition Board. He has been study director also for various studies dealing with food safety, including the landmark study on Scientific Criteria to Ensure Safe Food, and with the development of specifications for high-energy, nutrient-dense, emergency relief rations. He has also contributed to the work of the Roundtable on Environmental Health Sciences, Research, and Medicine with the Board on Health Sciences Policy. Dr. Molins received his Ph.D. in food science from Iowa State University, where he was later assistant and associate professor of food microbiology. He has worked on agro-industrial development for the United Nations Industrial Development Organization in Central America and for the Joint Division of the Food and Agriculture Organization of the United Nations/International Atomic Energy Agency, based in Austria, on food irradiation and food safety projects and research programs in Asia, Africa, the Middle East, and Latin America. He is the author of 46 scientific papers, 3 books, and numerous abstracts and has offered conferences on food safety in 24 countries.

ANDREW M. POPE, Ph.D., is director of the Board on Health Sciences Policy in the Institute of Medicine. With a Ph.D. in physiology and biochemistry, his primary interests are in science policy, biomedical ethics, and the environmental and occupational influences on human health. During his tenure at the National Academies and since 1989 at the Institute of Medicine, Dr. Pope has directed numerous studies on topics that range from injury control, disability prevention, and biologic markers, to the protection of human subjects of research, NIH priority-setting processes, organ procurement and transplantation policy, and the role of science and technology in countering terrorism. Dr. Pope is the recipient of the NAS President's Special Achievement Award and the IOM's Cecil Award.

JUDITH L. ESTEP is a senior program assistant at the Institute of Medicine. She has worked at The National Academies/Institute of Medicine since 1986 and has provided administrative support for more than 35 published reports.

BENJAMIN HAMLIN received his bachelor's degree in biology from the College of Wooster in 1993 and a degree in health sciences from the University of Akron in 1996. He then worked as a surgeon's assistant in the fields of vascular, thoracic, and general surgery for several years before joining the National Academies in 2000. As a research assistant for the Division on Earth and Life Studies at the National Academies, Mr. Hamlin worked with the Board on Radiation Effects Research on projects studying the health effects of ionizing and non-ionizing radiations on the human body. He was also involved with the U.S. Bangladesh Advisory Council, an organization that promotes governmental cooperation between the United States and Bangladesh on matters of trade and health care. He served as a research assistant with the Board on Health Sciences Policy at the Institute of Medicine through April 2004.

ERIN MCCARVILLE joined the National Academies in 2003 as a project assistant for the Committee on Science, Engineering and Public Policy. She worked as the senior project assistant for the NASA project through July 2005. Ms. McCarville received her bachelor's degree in biology and public policy from Pomona College in Los Angeles, California. Before working at the academies, she conducted research on rural environmental health for the Center for Community Action and Environmental Justice in Los Angeles. She also worked as a teaching and research assistant for Pomona College's Animal Physiology Department, as a plant biology researcher for the Chicago Botanic Gardens, and as an intern for Senator Barbara Boxer.

VILIJA TEEL works as the senior project assistant for the Board on Health Sciences Policy. She joined the Bioastronautics Roadmap project in July 2005. Ms. Teel also provides support to the Committee on Ethical Considerations for Revisions to DHHS Regulations for Protection of Prisoners Involved in Research. Prior to joining IOM, she worked as a program assistant for research, evaluation and development in the School of Language Studies within the Foreign Service Institute. Ms. Teel earned a B.A. in English/linguistics from Vilnius University, Lithuania. In addition to English, she has a good grasp of many other languages.

LISA M. VANDEMARK, Ph.D., worked as a senior program officer for the NASA project until May 2005. Her work focused on helping NASA understand the risk associated with manned space flight. As a geographer, she acted as consultant on education, training, and capacity building to the Committee on Earth Observation Satellites (CEOS). She has a Ph.D. in geography from Rutgers University. Prior to joining the staff of the IOM, Dr. Vandemark was a staff officer at the Board on Earth Sciences and Resources of the National Research Council and a research associate at Rutgers University.

MELVIN H. WORTH, Jr., M.D., is a scholar-in-residence at the Institute of Medicine. Dr. Worth completed his surgery residency at New York University–Bellevue in 1961 and remained on that faculty for 18 years. He founded the Bellevue Trauma Service in 1966 and continued as director until 1979, when he left to become director of surgery at Staten Island University Hospital. He served for 15 years with the New York State Office of Professional Medical Conduct and 8 years as a member of the New York State Hospital Review and Planning Council (for which he was chair in 1993). He is a fellow of the American College of Surgeons, the American College of Gastroenterology, and the International Society for Surgery and holds memberships in the American Association for the Surgery of Trauma, the Society for Critical Care Medicine, the Association for Academic Surgery, the New York Surgical Society (of which he was president in 1979), and other academic and professional organizations. Dr. Worth retains his appointments at New York University and the Uniformed Services University of the Health Sciences and has served as a clinical professor of surgery at SUNY–Downstate Medical Center. He has served on two editorial boards and has authored one textbook and 60 original articles. Dr. Worth has participated in IOM studies on Fluid Resuscitation for Combat Casualties as senior adviser to the Committee on Creating a Vision for Space Medicine During Travel Beyond Earth Orbit and the Longitudinal Study of Astronaut Health.

Index

A

Acceptable risk, determination of, 67
Actuarial data, using to estimate likelihood of intrinsic health alterations, 2, 11, 15, 49–50
Advanced human life support, the case of, 82–83
Advanced Human Support Technologies, 10, 82
Agency for Healthcare Research and Quality (AHRQ), 62
  National Guideline Clearinghouse, 62
AHRQ. *See* Agency for Healthcare Research and Quality
Airway management, operational issues in, 52
Analog environments, 32
Anion exchange resin filtration, 29
Apollo program, 73
Appetite, lack of, 38
Approximate reasoning, 63
Astronaut Office, 11, 42, 61, 64–65
Autonomous medical care, 51–54
  biological issues, 51
  health care delivery issues, 51
  operational issues, 51

B

Bayesian sequential trials approach, 79, 122
  using to address the small sample size, 13, 15, 78–80
Bayesian updating, 8, 13, 69, 120–123
BCPR. *See* Bioastronautics Critical Path Roadmap
Behavioral health and performance, priority of, 65
Behavioral health risks
  addressing issues of human sexuality in long-duration missions, 2
  grouping into categories based on clinical outcomes, 2, 11, 15, 42–44
  mitigation plans for, 32
  reclassification of, 41–44
Behavioral interventions, for mental disorders, 44
Behavioral research support, conducting periodic assessment of additional risks from lack of resources and using this to make decisions about, 13, 15, 75–78

Bioastronautics Critical Path Roadmap (BCPR), 17, 66
Bioastronautics Roadmap (BR)
- current printed and on-line versions of, 63, 68
- as a dynamic and current database, 2, 9, 14, 68–70
- hyperlinks in "Important References" section, 68
- overall assessment of, 23–24
- Risk Rating Categories and Priority Definitions, 20
- updating and maintaining, 24

Bioastronautics Roadmap (BR) content, 25–59
- creating the cross-cutting category "Food and Nutrition," 37–40
- creating the cross-cutting category "Human Systems Integration," 35–37
- fusing the relationships between human factors and technology in, 34–40
- overarching issues, 26–34
- specific issues, 40–54

Bioastronautics Roadmap (BR) context, 73–85
- addressing the challenges posed by the small sample size, 78–80
- analysis and prioritization to meet the launch schedule, 75–78
- the case of advanced human life support, 82–83
- efficiency and technology issues, 80–82
- organizational characteristics and risk, 73–75

Bioastronautics Roadmap (BR) process, 60–72
- determination of acceptable risk, 67
- Independent Health and Medical Authority, 70–71
- keeping the BR current, 69–70
- risk assessment, 60–66
- risk communication, 68–69

Bioastronautics Science Management Team (BSMT), 19, 61, 64–65
Biological issues, for autonomous medical care and self-care, 51
Bleeding control, operational issues in, 52
BMAD. *See* Bone mineral apparent density
BMD. *See* Bone mineral density
Bone fracture risk, associated with prolonged exposure to microgravity, 115–119
Bone loss
- countermeasures to, 32
- priority of, 65

Bone mineral apparent density (BMAD), 116
Bone mineral density (BMD), 115–116
BR. *See* Bioastronautics Roadmap
British Medical Research Council Cancer Trials, 121
BSMT. *See* Bioastronautics Science Management Team
Bush, George W., 20, 22

C

CAIB. *See* Columbia Accident Investigation Board
Cardiopulmonary resuscitation (CPR), operational issues in, 52
Central nervous system (CNS) impairment, 6, 46–47, 49
CEV. *See* Crew Exploration Vehicle
Challenger Space Shuttle, 73–74
Chief Engineer, 70
Circadian rhythm, problems with, 41
Clinically acute problems, 53
CNS. *See* Central nervous system impairment
Cochrane Collaboration, 62
Cognitive therapy, brief forms of, 43
Columbia Accident Investigation Board (CAIB), 70, 74

Columbia Space Shuttle, 73
Committee on Aerospace Medicine and Medicine in Extreme Environments, 90
Committee on Review of NASA's Bioastronautics Roadmap, 4, 90–95
charge to, 21–22
Content
recommendations on overarching issues in the BR, 14
recommendations on specific issues in the BR, 11–12, 15
Context, recommendations on specific issues in the BR, 12–15
Continuous Risk Management Program, 8, 67–69
Coping skills, 37
affecting others' work productivity, 29
Cost–utility analysis, 52
Countermeasure Readiness Level (CRL), 6, 29–31
status of, 51, 97–114
Countermeasures, 29
accelerating development of, 1, 6, 14, 29–34, 115
cost-effective, 61
for mental disorders, 43
CPR. *See* Cardiopulmonary resuscitation
Crew Exploration Vehicle (CEV), 76
Crew selection criteria, validating current and future, 2, 11, 15, 41–42
Crews
cognitive capabilities not matching task demands, 41
increased task performance expectations, 35
international, increasing complexity of health care delivery, 54
morale of, 38
CRL. *See* Countermeasure Readiness Level
Cross-cutting categories, 2, 10, 35–40
Cultural preferences and expectations, 40

D

Data-based evidence, in decision making, 8, 120–123
Decision making, integration of data-based evidence and expert opinion in, 8, 120–123
Bayesian updating, 8, 13, 69, 120–123
Deoxyribonucleic acid (DNA), damage to, 46
Depression
major mood swings related to, 46
nutritional factors in understanding, 28
Derivation phase, 77
Design Reference Missions, 1, 3, 5, 11, 17–19, 21, 25, 49–50, 67, 69, 74–76, 79n, 83
Digital simulation, 32, 70
Disaggregation, of risk from mitigation of risk, 62–63
DNA. *See* Deoxyribonucleic acid
Dual energy X-ray absorptiometry (DXA), 115–117

E

"Efficiency" risks, 80–82
EVA. *See* Extravehicular activity
Evidence-based interventions, for mental disorders, 43
Evidence-based risk assessment, 62
"Expert opinion," in decision making, 8, 69, 120–123
Extravehicular activity (EVA), 6, 83
clothing for, 36, 49
decompression procedures for, 76
technology for, 82

F

Fatigue, cognitive effects associated with, 45

Findings and recommendations, 5–9
radiation effects, establishing risk-specific radiation exposure levels, 6–7
risk assessment, 7–8
risk communication and keeping the BR current, 8–9
status of readiness levels, 5–6
Fixed effects methods, using to address the small sample size, 13, 15, 78–80
Food, grown and processed during flight, 39
"Food and Nutrition" category, creating, 2, 10, 14, 34–40
Fracture risk, assessment of, 116–117
Fuzzy sets, 63

G

"*G*-suit" system, 63
Galactic cosmic rays (GCRs), 46–48
Gamma rays, 47
GCRs. *See* Galactic cosmic rays
Ground-crew relations, dysfunctional, 35–36
Group relations, effective, 44

H

Habitability studies, 36
Hazardous environments of space, 80
Health care delivery issues
for autonomous medical care and self-care, 51
impact of crew selection on, 54
in nonterrestrial environments, 52
Health risks, quantitatively evaluating, 2
Hierarchical random methods, using to address the small sample size, 13, 15, 78–80
High-mass, high-*Z,* high-energy radiation (HZE), 6, 47–48
Homeostasis, 38
Homogeneous teams, 77
Houston Workshop, on the Bioastronautics Roadmap, 65, 79n, 89–95
Human life support, the case of advanced, 82–83
Human performance failure, due to organizational and cultural factors, 12–13, 15, 73–75
Human sexuality, in long-duration missions, addressing issues of, 11, 15, 44
Human system, impossibility of reengineering, 81
"Human Systems Integration" category, creating, 2, 10, 14, 34–40
Human-robot interactions, 36
Human-technology interactions, 35
Hyperlinks, in "Important References" section of the BR, 68
HZE. *See* High-mass, high-*Z,* high-energy radiation

I

ICRP. *See* International Commission on Radiological Protection
Independent Health and Medical Authority (IHMA), 70–71
establishing, 12, 15, 70–71
Independent Technical Authority (ITA), 70–71
Individual tolerance, 47
Inherent risks, 29
Institute of Medicine (IOM), 3, 13, 21, 64, 80
International Commission on Radiological Protection (ICRP), 47
International crews, increasing complexity of health care delivery, 54
International Space Station (ISS), 3, 21, 34, 74. *See also* Design Reference Missions
investigations aboard, 32, 77

Interpersonal therapy, brief forms of, 43
Interventions for mental disorders
  behavioral, 44
  evidence-based, 43
Intrapersonal factors, 41
Intrinsic health alterations, using actuarial data to estimate likelihood of, 2, 11, 15, 49–50
Iodine, used as a bacteriostatic agent, 28, 33
IOM. *See* Institute of Medicine
Ionizing radiation, from the solar wind, 46
Iran Nonproliferation Act, 76
Isolation studies, 45
ISS. *See* International Space Station
Issues. *See* Overarching issues in the BR content; Specific issues
ITA. *See* Independent Technical Authority

J

Johnson Space Center, 50, 89

L

Launch schedule, analysis and prioritization to meet, 75–78
LEO. *See* Low Earth orbit
Long-duration space flights, 35, 41, 46
  addressing issues of human sexuality in, 11, 15, 44
  assessing the impact on crew's health and incremental risk, 49–50
  simulating, 77
Longitudinal Study of Astronaut Health, 50
Low Earth orbit (LEO), 76
  space travel below, 47
Low-readiness areas, 32
Lunar surface sojourn. *See* Design Reference Missions

M

Magnetic resonance (MR) scans, 53
Mania, major mood swings related to, 46
Mars journey. *See* Design Reference Missions
Mealtimes, social interactions and connotations of, 40
Medical care
  autonomous, 51–54
  priority of, 65
Mental disorders, 42
  interventions or countermeasures for, 43
  nutritional factors in understanding, 28
Mental health risks
  affecting crew health and mission success, developing a system for quantitative evaluation of, 12, 15, 51–54
  quantitatively evaluating, 2
Methodology, 22–23, 89–95
  Committee on Aerospace Medicine and Medicine in Extreme Environments, 90
  Committee on Review of NASA's Bioastronautics, 90–95
  Institute of Medicine workshops, 90–95
Microgravity support, conducting periodic assessment of additional risks from lack of resources and using this to make decisions about, 13, 15, 75–78
Mir Space Station, 29, 41
Mitigation strategies, 27, 32
Mobility aids, 36
Mood swings, major, 45
Moon surface sojourn. *See* Design Reference Missions
MR. *See* Magnetic resonance scans
Muscle loss, 39
Musculoskeletal weakness, 36

N

Narcissism, 42
Narrative descriptions of risks, 63
NASA. *See* National Aeronautics and Space Administration
National Aeronautics and Space Administration (NASA), 2–8, 10–11, 13, 17–19, 60, 73, 115, 120. *See also* Bioastronautics Roadmap (BR)
  challenges to managers of, 25
  constraints on the ISS imposed by, 76
  Continuous Risk Management Program, 8, 67–69
  Independent Technical Authority, 70–71
  Office of Biological and Physical Research, 18
  Office of Space Flight, 18
  Office of the Chief Health and Medical Officer, 18, 61, 64
National Guideline Clearinghouse, 62
National Space Biomedical Research Institute (NSBRI), 19, 60
Naval Submarine Medical Research Laboratory (NSMRL), 33
Naval Undersea Medical Institute (NUMI), 33
Neurobehavioral problems, 41
Neurovestibular functionality, 27
  priority of, 65
NSBRI. *See* National Space Biomedical Research Institute
NSMRL. *See* Naval Submarine Medical Research Laboratory
NUMI. *See* Naval Undersea Medical Institute
Nutritional factors, 38, 118
  in understanding depression and other mental disorders, 28

O

Office of Biological and Physical Research, 18
Office of Space Flight, 18
Office of the Chief Health and Medical Officer, 18, 61, 64
Operational issues and solutions, 5
  airway management, 52
  for autonomous medical care and self-care, 51
  bleeding control, 52
  cardiopulmonary resuscitation, 52
  suction, 52
Organizational characteristics, and risk, 73–75
Outcome-oriented approach, 31
Overarching issues in the BR content, 9–10, 26–34
  countermeasure and technology readiness levels, 29–31
  creating the cross-cutting categories "Human System Integration" and "Food and Nutrition," 2, 10
  interactions among risks, 10, 27–29
  status of readiness levels, 29–34
  the time factor and its impact on risk, 9, 26–27

P

Palatability, 40
Performance. *See also* Human performance failure
  increased task expectations of crew, 35
  priority of, 65
  psychological and physical impacts of space flight on, 44–46
Personality disorders, 42
Physical health risks
  affecting crew health and mission success, developing a system for quantitative evaluation of, 12, 15, 51–54
  quantitatively evaluating, 2
Physical restraints, 36
Possibility theory, 63
Preliminary Considerations Regarding NASA's Bioastronautics Critical Path Roadmap, 21

President's space initiative, 18, 22–23, 26, 74–75
vision for U.S. space exploration, 22
Principal recommendations, 5–9, 14
accelerating countermeasure and technology development, 6, 14, 29–34
ensuring that the BR is a dynamic and current database that enhances communication and conveys information effectively, 9, 14, 68–70
establishing a safe radiation exposure level for all relevant risks, 7, 14, 46–49
incorporating quality-of-evidence measures, 8, 14, 61–62
representing risk severity separately from the state of the mitigation strategy or countermeasure, 8, 14, 62–63
using standard uncertainty analysis techniques to quantify risk uncertainty, 8, 14, 63–66
Prioritization, 65
threshold for, 5
Process, recommendations on specific issues in the BR, 12, 15
Psychosocial adaptation, 41

Q

Quality-of-evidence measures, incorporating, 2, 8, 14, 61–62
Quantitative computed tomographic (QCD)-derived measurements, 116

R

Radiation exposure
mitigation plans for, 32
priority of, 65
Radiation exposure levels
establishing risk-specific, 6–7, 46–49
establishing safe, 1–2, 7, 14, 46–49
increased in space, 39
Random methods, hierarchical, used to address the small sample size, 13, 15, 78–80
RBCs. *See* Red blood cells
Readiness levels. *See* Countermeasure Readiness Level (CRL); Technology Readiness Levels (TRLs)
Reclassification, of behavioral health risks, 41–44
Recommendations, 14–15
principal, 14
summary of, 14
Recommendations on overarching issues in the BR content, 14
creating two new cross-cutting categories: human systems integration and food and nutrition, 10, 14, 34–40
defining, quantifying, and mitigating interactions among risks, 10, 14, 27–29
labeling risks by relevance to operational requirements and temporal urgency, 9, 14, 26–27
Recommendations on specific issues in the BR content, 11–12, 15
addressing issues of human sexuality in long-duration missions, 11, 15, 44
developing a system for quantitative evaluation of mental and physical health risks affecting crew health and mission success, 12, 15, 51–54
grouping behavioral health risks into categories based on clinical outcomes, 11, 15, 42–44
using actuarial data to estimate the likelihood of intrinsic health alterations as part of the selection criteria for the Mars mission crew, 11, 15, 49–50

validating current and future crew selection criteria, 11, 15, 41–42
Recommendations on specific issues in the BR context, 12–15
conducting periodic assessment of additional risks from lack of resources and using this to make decisions about microgravity and behavioral research support, 13, 15, 75–78
human performance failure due to organizational and cultural factors, 12–13, 15, 73–75
reframing risks as either health or technology related, 14–15, 80–82
using Bayesian sequential trials approach and hierarchical random or fixed effects methods to address the small sample size, 13, 15, 78–80
Recommendations on specific issues in the BR process, 12, 15
establishing an independent health and medical authority, 12, 15, 70–71
Red blood cells (RBCs), decreased release rate of, 39
Reengineering, 81
"Related risks," 28
"Retiring" risks, 63
Risk assessment, 7–8, 19–20, 60–66
disaggregation of risk from mitigation of risk, 62–63
evidence-based, 62
expressing uncertainty, 63–66
Risk communication, 68–69
and keeping the BR current, 8–9
Risk uncertainty, using standard uncertainty analysis techniques to quantify, 8, 14, 63–66
Risks
defining, quantifying, and mitigating interactions among, 10, 14, 27–29
"efficiency," 80–81
identification of, 19
inherent, 29
labeling by relevance to operational requirements and temporal urgency, 9, 14, 26–27
narrative descriptions of, 63
organizational characteristics of, 73–75
quantitatively evaluating, 2
reframing as either health or technology related, 14–15, 80–82
representing severity of separately from the state of the mitigation strategy or countermeasure, 8, 14, 62–63
"retiring," 63

S

Safe Passage: Astronaut Care for Exploration Missions, 64
"Select-in" and "select-out" procedures, 41–42, 77
Self-care, 51–54
biological issues, 51
health care delivery issues, 51
operational issues, 51
Sex. *See* Human sexuality
Silver nitrate, using as a bacteriostatic agent, 28–29
Sleep loss, 41
Small Clinical Trials: Issues and Challenges, 13, 80
Small sample size, addressing the challenges posed by, 78–80
Social interactions, 40
Solar particle events (SPEs), 46, 48
Solar wind, ionizing radiation from, 46
Space Medicine Office, 65
Spacelab Life Sciences Shuttle missions, 39
Spatial orientation, maintaining, 36
Specific issues, 10–15, 40–54
assessing the sources and impact of long-duration space flight on crew's health and incremental risk, 49–50

autonomous medical care and self-care, 51–54
psychological and physical impacts of space flight on performance, 44–46
radiation effects and establishing risk-specific radiation exposure levels, 46–49
reclassification of behavioral health risks, 41–44
recommendations related to specific issues in the BR content, 11–12
recommendations related to specific issues in the BR context, 12–14
recommendations related to specific issues in the BR process, 12
SPEs. *See* Solar particle events
Stakeholders, 9, 121–122
Standard uncertainty analysis techniques, using to quantify risk uncertainty, 8, 14, 63–66
Status of readiness levels, 5–6, 29–34, 97–114
Stress
analyzing responses to, 46
high in tight quarters, 32
Stress hormones, increased concentrations of, 45
Submarine studies, 33
Suction, operational issues in, 52
Surrogates, using in assessment, 32

T

Teams. *See also* Crews
dynamics of, 77
homogeneous, 77
Technology development, 80–82
accelerating, 1, 6, 14, 29–34
Technology Readiness Levels (TRLs), 6, 13, 29–31, 78, 83
status of, 97–114
testing operationally, 76
Terrestrial parameters, 83
Thyroid dysfunction, 33
following water contamination, 28
Time factor, and its impact on risk, 9, 26–27
TRLs. *See* Technology Readiness Levels

U

Ultrasound units, handheld diagnostic, 53
Uncertainty
analysis techniques for, 8, 14, 63–66
expressing, 63–66
Unranked areas, 32
U.S. Department of Defense, 10

V

Variety issues, in diet, 40
Vegetarian diet, 40
Vitamin status, 38

W

Water contamination, and consequent thyroid dysfunction, 28
Water quality monitoring, mitigation plans for, 32
WHO. *See* World Health Organization
Women crew members, 34
Work productivity, of others, coping skills affecting, 29
Workshop on the Bioastronautics Roadmap, 65, 79n, 89–95
results of the CB/SD/BSMT consensus, 66
World Health Organization (WHO), 115